Mesomorphic Order in Polymers

and Polymerization in Liquid Crystalline Media

Alexandre Blumstein, EDITOR
University of Lowell

A symposium sponsored by the Division of Polymer Chemistry, Inc. at the 174th Meeting of the American Chemical Society Chicago, Ill., August 29–30, 1977

ACS SYMPOSIUM SERIES **74**

AMERICAN CHEMICAL SOCIETY
WASHINGTON, D. C. 1978

Library of Congress CIP Data

Mesomorphic order in polymers and polymerization in liquid crystalline media.
(ACS symposium series; 74)

Includes bibliographical references and index.

1. Polymers and polymerization—Congresses. 2. Liquid crystals—Congresses.
I. Blumstein, Alexandre. I. American Chemical Society. Division of Polymer Chemistry. III. Series: American Chemical Society. ACS symposium series; 74.

QD380.M47 547'.84 78-9470
ISBN 0-8412-0419-5 ACSMC8 74 1-264 1978

PRINTED IN THE UNITED STATES OF AMERICA

ACS Symposium Series

Robert F. Gould, *Editor*

FOREWORD

The ACS SYMPOSIUM SERIES was founded in 1974 to provide a medium for publishing symposia quickly in book form. The format of the SERIES parallels that of the continuing ADVANCES IN CHEMISTRY SERIES except that in order to save time the papers are not typeset but are reproduced as they are submitted by the authors in camera-ready form. As a further means of saving time, the papers are not edited or reviewed except by the symposium chairman, who becomes editor of the book. Papers published in the ACS SYMPOSIUM SERIES are original contributions not published elsewhere in whole or major part and include reports of research as well as reviews since symposia may embrace both types of presentation.

CONTENTS

PREFACE

Mesomorphic order in polymers is important to the field of high strength materials such as the ultra-high modulus fibers recently developed by industry from stiff chain polymers and carbon pitches. However, other fundamental aspects of polymer chemistry and physics are involved also. Mesomorphic order in polymers is tied to the problem of local segmental order within macromolecular coils. It has a direct bearing on the understanding of the complex morphology of crystalline polymers in the regions of intermediate order. The relationship between crystallinity, molecular order, and stereoregularity and problems relevant to biophysics of macromolecules, such as intramolecular crystallization, formation of tertiary structures, and formation of liquid crystalline morphoses in living tissues and cell organelles, are some of the topics which are discussed in this volume.

Whereas the subject of stiff chain polymers has received much attention in two recent ACS Meetings (Witco Symposium honoring Dr. Paul Morgan held in New York in April 1966 and symposium on Rigid Chain Polymers held in New Orleans in March 1977), the wider aspects of mesomorphic order in polymers have been discussed only occasionally at meetings. The chapters assembled in this volume constitute the proceedings of the first symposium devoted to broader aspects of this rapidly developing field and, as such, provide a valuable source of information to the student and scientist.

University of Lowell
Lowell, MA 01854
April 11, 1978

ALEXANDRE BLUMSTEIN

1

X-Ray Diffraction from Polymers with Mesomorphic Order

S. B. CLOUGH and A. BLUMSTEIN
Polymer Program, Department of Chemistry, University of Lowell, Lowell, MA 01854
A. deVRIES
Liquid Crystal Institute, Kent State University, Kent, OH 44242

Polymers with side chain structure similar to that of low molecular weight liquid crystalline compounds can achieve various levels of organization in the bulk. These polymers are sometimes formed by polymerization of vinyl monomers that themselves exhibit mesomorphic behavior. In other cases, they can be obtained from monomers that do not form liquid crystalline states. At one extreme the structure of the polymer is highly organized, approaching that of crystalline polymers and giving rise to a number of x-ray diffraction peaks. At the other extreme the polymer chains are disorganized, with x-ray diffraction patterns that resemble those from amorphous polymers. The relation between the molecular structure and the mesomorphic order has been previously discussed (1). In this paper, the x-ray diffraction results for a number of polymers with stiff and/or bulky side groups are discussed.

X-ray diffraction photographs from unoriented liquid crystalline states of low molecular weight compounds are characterized by an inner ring (at 2θ = 2-5 degrees) related to the length of the molecule, and an outer ring (20$\sim$20 deg). The former is sharp for smectic states where the molecules are mutually parallel and arranged in planes. The spacing calculated (Bragg equation) is the distance between these planes. The inner ring is diffuse in the case of nematic states, where the molecules in a volume element lie approximately parallel to each other but are not organized into a lamellar structure. The sharpness of the outer ring(s) depends principally on the extent of lateral packing between molecules. In many cases a diffuse halo near 2θ = 20 deg shows the absence of long range lateral order between the molecules.

0-8412-0419-5/78/47-074-001$05.00/0

A number of different smectic states exist among the low molecular weight mesomorphic compounds. The molecules, for example, might be normal to the planes or tilted relative to the normal. The different states have been given various letter designations (Smectic A, B, C etc.) and deVries (2) has classified the x-ray patterns obtained from the various types of smectic (and also nematic) states.

The patterns obtained from the polymers discussed here show features similar to those of low molecular weight compounds. The inner ring is now related to the organization and length of the polymer side groups, while the outer rings depend on the lateral packing between the side groups. X-ray patterns and polymer organization are discussed here in terms of the classification of deVries. This does not necessarily imply miscibility with a low molecular weight liquid crystal of the same letter designation.

Indeed, the polymers discussed here are not fluid materials but are hard and brittle, and are not liquid-crystals in the usual sense of this term. In most cases heating does not change the level of organization up to temperatures where the polymer starts to decompose. Other comparisons and distinctions will be drawn between these polymers and low molecular weight mesomorphic states below. Some comments will also be made on the organization of the polymers as compared to semi-crystalline polymers with small extent of crystallinity.

Experimental

The methods of preparation of the monomers are given in the references in Tables I-III.

X-ray diffraction data were obtained with a Warhus vacuum camera, and Rigaku wide angle (model SG-7B) and small angle diffractometers, all with Cu Kα radiation. A Leitz Ortholux polarizing microscope with a Mettler FP52 hot stage was used for examination of birefringent films of the polymers and for determining monomer transition temperatures. A Perkin Elmer DSC-1B differential scanning calorimeter was also used to obtain the transition temperature of monomers.

Results and Discussion

The polymers are presented in order of decreasing levels of organization, starting with highly ordered smectic and proceeding to poorly organized nematic states.

Smectic. Table I lists the monomers of some polymers that show a sharp inner ring and thus have smectic organization.

Poly(acryloyloxybenzoic acid), (PABA). When the corresponding monomer is polymerized in the melt to high degrees of conversion,* x-ray diffraction from the polymer shows a sharp ring corresponding to 17.7Å, and three sharp outer rings in addition to the outer halo (3). The spacing of the inner ring is less than twice the length of the side groups, suggesting that the latter are strongly tilted with respect to the smectic planes. We were unable to orient the sample to confirm this tilting. In the smectic state of a low molecular weight analog, nonyloxybenzoic acid, the molecules are tilted at approximately 45 degrees to the plane normal (4). The x-ray pattern most closely resembles that of the Smectic E_t (2).

The three outer rings show that there exists some regular packing between the side groups. It might be noted that for polymers we would expect to observe the halo, even if there are sharp outer rings present. This would be due to poorly organized regions. The ordered domains are not expected to extend uniformly throughout the specimen. The presence of the halo is similar to the case of semi-crystalline polymers.

Poly(methacryloyloxybenzoic acid), (PMBA). Films of this polymer cast from dimethylformamide give x-ray patterns with a sharp inner and a sharp outer ring, (5) in addition to the halo. This pattern resembles the pattern of Smectic B_t (2).

It is interesting to remark that in both abovementioned cases, the monomers are not mesomorphic.

Poly(N-p-methacryloyloxybenzylidene-p-aminobenzoic acid), (PMBABA). Polymerization at 213°C from the nematic state gives a polymer with a smectic structure. The spacing calculated from the sharp inner ring is approximately twice the length of the side chains (6). This suggests Smectic A organization with the side groups normal to the planes and the polymer main chain contained within the planes. Polymerization of the monomer from the nematic state in a 10kOe magnetic field gave an oriented specimen. X-ray

*Polymerization to low degrees of conversion and casting of films from a dimethylformamide solution gives a semi-crystalline polymer with up to eight x-ray diffraction lines.

Table I. Polymers with Smectic Organization

No.	Abbreviation of Polymer Structure of Monomer	Monomer Transitions, degree C	Polymer X-ray Data d, Å	D, Å	Ref.
1.	PABA CH_2=CHCOO-⟨Ph⟩-COOH	K 201 I	17.7		3
2.	PMBA CH_2=C(CH_3)COO-⟨Ph⟩-COOH	K 182 I	19.0		3,5
3.	PMBABA CH_2=C(CH_3)COO-⟨Ph⟩-CH=N-⟨Ph⟩-COOH	K_1 182 K_2 201 S 206 N	32.0	5.08	6,11
4.	PChMA CH_2=C(CH_3)COO[Chol]	K 114.8 I (111.8C)	35.3	6.35	8
5.	PChAB CH_2=CHCOO-⟨Ph⟩-COO[Chol]	K 128 C Poly'n	45.2	6.08	8
6.	PdiABAB (CH_2=CHCOO-⟨Ph⟩-CH=N$)_2$-⟨Ph⟩-		22.2	5.0	9

diffraction shows the outer halo to be positioned at 90° to the arcs diffracted from the planes. Figure 1 shows the x-ray pattern from an oriented specimen. This confirms the similarity of this organization with the Smectic A. No regular packing exists between the side chains. The average distance between the side chains is D=5.08Å calculated from $1.117\lambda = 2D\sin\theta$ (7). A schematic drawing of the structure is shown in Figure 2.

An unusual feature of Figure 1 is the multiplicity of higher orders of diffraction from the smectic planes. The following orders have been observed: 1, 2, 3, 5, 6, 7, 9, 10, 11 and 13. A few higher order peaks have been reported for polymers and low molecular weight compounds with smectic organization. The polymer PMBABA must have a high degree of positional regularity along the side group, yet no ordered lateral packing of the side groups.

Poly(cholesteryl methacrylate), (PChMA) and Poly(cholesteryl p-acryloyloxybenzoate), (PChAB). The structure of these polymers is classified as Smectic A (8). The interplanar spacing is approximately twice the length of the side group. The structure is discussed in some detail in another paper in this volume.

Figure 3 shows the effect of different methods of specimen preparation on the sharpness of x-ray lines from smectic planes. The intense, sharp peak in Figure 3a is from the polymer PChMA prepared by bulk polymerization of the isotropic monomer. Dissolving the sample in benzene and subsequent film casting gives a poorly organized structure, (Figure 3b). Subsequent annealing of this specimen at 180°C (Tg~160°C) in a vacuum oven for two hours allowed reorganization of the chains resulting in the diffraction shown in Figure 3c. In other cases in which smectic organization is achieved during polymerization of the monomer directly, rapid precipitation of the polymer from solution gives a powder which shows only a very diffuse halo at the scattering angle of the inner ring. In discussing the mesomorphic structure of polymers, it is important that the method of sample preparation be known.

Poly(di(N-p-acryloyloxybenzylidene)-p-diaminobenzene), (PdiABAB). The polymerization of this difunctional monomer and the copolymerization with a monofunctional monomer have been previously described, (9). Polymerization of this monomer in its nematic state in a magnetic field of 4,000 Oe leads to an oriented polymer. The diffraction pattern of this poly-

Figure 1. X-ray diffraction from PMBABA showing orientation of the smectic planes. The 32Å line is partially hidden by the beam stop.

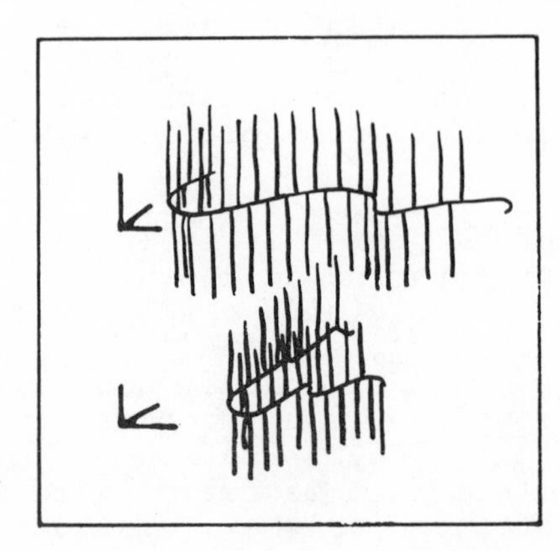

Figure 2. Schematic diagram of the smectic A structure in PMBABA

mer is analogous to a pattern of a Smectic A mesophase.

Intermediate Organization. The two monomers listed in Table II give rather intense inner rings that are broader than those from the lamellar states described above. Figure 4 shows the diffraction from poly(N-p-butoxybenzylidene-p-aminostyrene),(PBBAS). The Bragg spacing calculated from the inner ring is less than twice the length of the side chain (with extended n-alkyl groups). Further, careful measurement of the angles of the first and second order peaks, shows (sin θ) second order $<$ 2(sin θ) first order. These features indicate a structure which differs from a well organized smectic. Further, the high intensity of the inner ring implies a state more organized than nematic.

We have been unable to fit these data into the classification (for low molecular weight compounds) of deVries. The broadening of the lines may come from distortions of a smectic organization with interpenetration of the n-alkoxy chains from different molecules. The outer diffuse halo shows that no lateral regular packing exists between the side chains.

Nematic. Table III gives three monomers that polymerize to states that we have classified as nematic. Since some of the properties of these polymers differ widely from those of low molecular weight nematic states, the experimental results should be closely examined.

Poly(N-(p-cyanobenzylidene)-p-aminostyrene), (PCBAS). Polymerization of the monomer from the nematic and isotropic states proceeds over a few hour period (10). Samples polymerized at 127°C for 24 hours show a very weak diffuse inner ring at $2\theta \sim 4.2$ degrees. This is much less intense than expected from comparisons with low molecular weight nematics. Within the outer halo there is a sharp line corresponding to 4.43Å. Films of the polymer are transparent, but are weakly birefringent. Light of low intensity is transmitted between crossed polarizers in the optical microscope.

Based on comparisons with low molecular weight compounds, the optical clarity and the weak inner ring suggest that the organization is isotropic. However, nematic order was assigned to the polymer based on observation of optical anisotropy (10). The sharp line of medium intensity within the outer halo suggests some parallel packing of the side chains. The spacing is the same as that found for other side

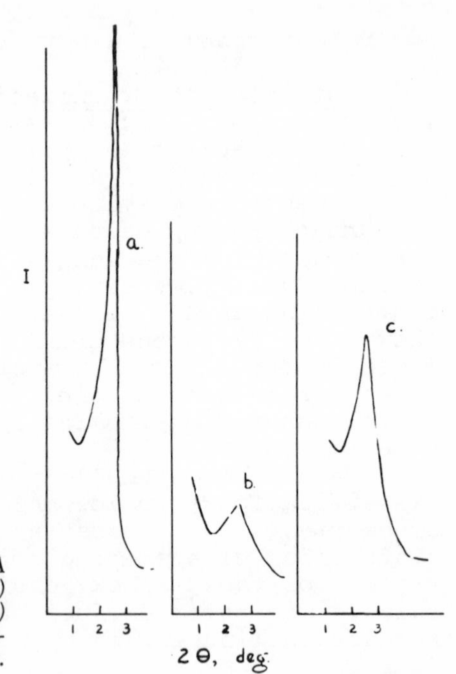

Figure 3. X-ray diffraction from PChMA prepared under different conditions. (a) Polymerized from the isotropic melt; (b) films cast from benzene solution; (c) specimen from (b) annealed at 180°C for 2 hr.

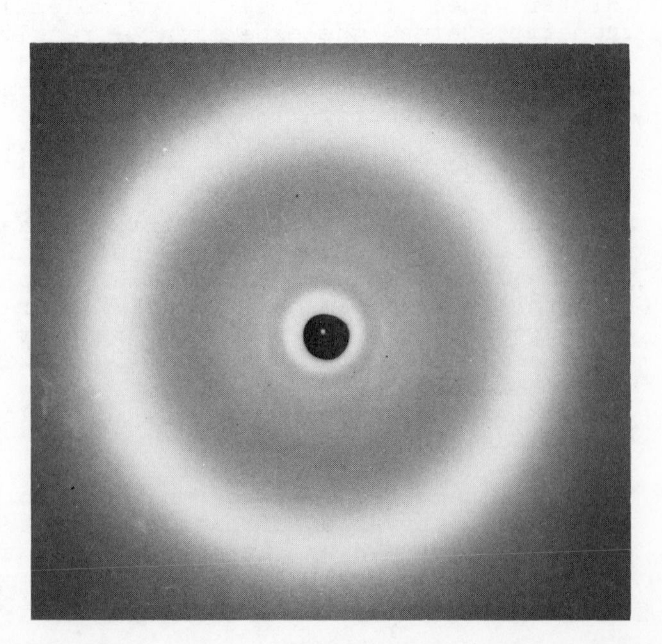

Figure 4. X-ray diffraction from PBBAS

Table II. Polymers with Intermediate Organization

No.	Abbreviation of Polymer Structure of Monomer	Monomer Transitions, degrees C	Polymer X-ray Inner Ring	Data D,Å	Ref.
1.	PBBAS $CH_2=CH-C_6H_4-N=CH-C_6H_4-OC_4H_9$	K 88.3 N 120.6 I	$2\theta=3.1^\circ$	4.95	5,11
2.	PHBAS $CH_2=CH-C_6H_4-N=CH-C_6H_4-OC_6H_{13}$	K 94 S 97.5 N 116 I	$2\theta=3.3^\circ$	5.00	11

Table III. Polymers with Nematic Organization

No.	Abbreviation of Polymer Structure of Monomer	Monomer Transitions, degrees C	Polymer X-ray Inner Ring	Data D,Å	Ref.
1.	PCBAS $CH_2=CH-C_6H_4-N=CH-C_6H_4-C\equiv N$	K 113.8 N 140.5 I	4.2° very weak	4.9	10
2.	PPMAS $CH_2=CH-C_6H_4-N=CH-C_6H_4-CH=N-C_6H_4-CH=CH_2$	K 180 N Poly'n	---	5.1	9
3.	PdiABH $CH_2=CHCOO-C_6H_4-CH=N-N=CH-C_6H_4-OOCCH=CH_2$	K 140 N Poly'n	---	5.3	9

chains containing the Schiff base structure and phenyl rings when regular packing is achieved (9). Further, this sharp peak is absent from the x-ray photographs from samples precipitated from solution. This latter method of sample preparation leads to a less ordered structure than bulk polymerization. Thus, the sharp line must be due to parallel packing between side groups, not other intramolecular effects.

The low intensity of the x-ray inner ring could be due to two factors. 1) The nematic organization might not be extensive. As mentioned above, the ordered domains may not extend continuously throughout the sample in polymers, but would be interrupted by disordered regions. 2) The diffuse inner ring for nematic states in small molecules is caused by the low electron density at the ends of the molecules. In the case of polymers with nematic packing of the side chains, one end of the nematogenic structure is bonded to the polymer main chain with no abrupt decrease in density. Thus, there are fewer electron density fluctuations per unit volume. To our knowledge, there are no reported cases where polymers with nematic organization of side groups give an inner ring of moderate intensity.

The turbidity of the nematic states of low molecular weight compounds is due primarily to fluctuations in the orientation of the director over distances on the order of the wave length of visible light (12). The optical clarity of PCBAS indicates that the regions of nematic organization are much smaller in size. Light is not appreciably scattered, nor are the typical nematic textures resolvable in the polarizing microscope.

Poly(p-phenylene-bis(N-methylene-p-aminostyrene)), (PPMAS), Poly(di(N-p-acryloyloxybenzylidene)hydrazine), (PdiABH). These difunctional monomers yield polymer networks upon bulk polymerization. No inner rings have been detected from the nematic organization. This has been explained (9) by the fact that both chain ends of the monomer are incorporated into the polymer main chains. In the case of complete conversion to polymer there are no side group ends giving large electron density fluctuations.

Comparison with Crystalline Polymers. The structure of the polymers discussed above differs from that in crystallizable polymers with low degrees of crystallinity (poorly developed structure). Here the tendency to organize comes from the parallel packing of the side groups. There is no evidence that polymer main chains are parallel to each other over long distances. In the smectic A and C states, we envision the polymer chain as coiled though confined to a plane. In the usual case of polymer crystals the main chains lie parallel in the crystalline regions.

Acknowledgement

Thanks are expressed to the NSF for partial support under the Grant DMR-75-17397.

Literature Cited

1. Blumstein, A., Macromol, (1977) 10, 872.
2. deVries, A., Pramana, (1975) 9, 93 Suppl. 1, and "Liquid Crystals," Chandrasekar, S. ed., Indian Acad. Sci., Bangalore, 1975.
3. Blumstein, A., Clough, S.B., Patel, L., Blumstein, R.B., and Hsu, E.C., Macromol., (1976) 9, 243.
4. Chystiakov, I., "Ordering and Structure of Liquid Crystals" in Adv. in Liquid Crystals, Vol. 1, Brown, G. ed., Academic Press, (1975).
5. Blumstein, A., Blumstein, R.B., Clough, S.B., and Hsu, E.C., Macromol, (1975) 8, 73.
6. Lim, L., Thesis in preparation for partial fulfillment of requirements for Ph.D. degree. Polymer Preprints, (1975) 16, No. 2, 241.
7. deVries, A., Mol. Cryst. Liq. Cryst., (1970) 10, 219.
8. Hsu, E.C., Clough, S.B., and Blumstein, A., Polym. Lett., (1977) 15, 545.
9. Clough, S.B., Blumstein, A., and Hsu, E.C., Macromol., (1976) 9, 123.
10. Hsu, E.C., and Blumstein, A., Polym. Lett., (1977) 15, 129.
11. Hsu, E.C., Lim, L.K., Blumstein, R.B., and Blumstein, A., Mol. Cryst. Liq. Cryst. (1976) 33, 35.
12. deGennes, P., The Physics of Liquid Crystals, Oxford University Press, London (1974).

RECEIVED December 8, 1977.

2

X-Ray Diffraction Studies on Mesomorphic Order in Polymers

J. H. WENDORFF, H. FINKELMANN, and H. RINGSDORF

Deutsches Kunststoff Institut, Darmstadt, and Institut für Organische Chemie, Universität Mainz, West Germany

In the past few years a considerable number of papers were published which were concerned with liquid crystalline structures in polymeric systems. Different routes were employed to obtain polymers with liquid crystalline structures or even thermodynamically stable liquid crystalline phases [1]. In general monomers containing mesogenic groups - groups which are known to have a tendency towards the formation of liquid crystalline structures, or rigid groups were used. Cases are known where the monomers exhibit liquid crystalline phases [2]. In that case the polymerization can be performed in anisotropic melts; frozen-in liquid crystalline structures and textures can be obtained in many instances [1,2]. In other cases the monomers do not display liquid crystalline phases. The formation of liquid crystalline polymer structures may nevertheless be possible due to the restriction of the motions of the individual repeat units [3].

Basically two different approaches have been used to obtain liquid crystalline structures in polymers: Either the mesogenic or rigid group was built into the polymer backbone or it was attached to the main chain as a side chain.

Main chain polymers. Due to the rigidity of the monomer units stiff polymer chains are obtained which according to the theory should have a tendency towards the formation of anisotropic melts [4]. The properties of these melts have been widely used to obtain fibers with very good mechanical properties [5]. Disadvantages of rigid chains are their high melting point and strong restrictions of reorientational motions which are necessary for inducing textural changes. One way of overcoming these disadvantages consists in putting

0-8412-0419-5/78/47-074-012$05.00/0

flexible spacer groups between mesogenic groups in order to decouple the orientational motions and the positional order of the individual repeat units [6]. It can be expected that in this case the mesogenic groups will form liquid crystalline structures in the same way as in the case of low molecular weight substances.

Side chain polymers. In the case of side chain polymers in which the mesogenic groups are attached to the backbone two ways of obtaining liquid crystalline structures are possible. If the side chain is rigidly attached to the backbone liquid crystalline side group structures can only be obtained if the chain conformation is distorted with respect to the chain conformation in the isotropic fluid state. An example will be discussed where the backbone is confined to a plane within smectic layers. If the side groups are decoupled from the backbone with respect to the positional and orientational order as well as with respect to the reorientational motions it is possible for the side groups to form liquid crystalline structures without distorting the conformation of the polymer main chain. One then expects that structure formation occurs in a way similar to that known in low molecular weight liquid crystals. Short mesogenic groups are expected to form nematic structures whereas long groups are expected to form smectic structures. These model considerations are discussed in detail in connection with the synthesis of the monomers and polymers in paper [7].

In this paper three kinds of polymers will be considered (polymers 1, 2 and 3) in which mesogenic groups are either attached to the flexible backbone via flexible spacer groups (polymers 1 and 2) or directly without spacer groups. The polymers studied are shown in table 1.

Experimental

The syntheses of the monomers and polymers based on model considerations are described in a separate paper [4]. The samples used in the experiments were obtained either by precipitating the polymers from solutions or by casting films from solutions. The films were studied "as received" and also after heating them up to temperatures above the melting point and cooling them down to room temperature.

The structure of the polymers was studied by means of X-ray diffraction. The experiments were performed

Table 1

Polymers studied, in most cases methacrylic backbones were used.

Polymer 1
Backbone-COO-$(CH_2)_n$-O-C_6H_4-COO-C_6H_5-R
$(CH_2)_n$ flexible spacer group, R variable

Polymer 2
Backbone-COO-C_6H_4-$(CH_2)_n$-COO-Cholesteryl
$(CH_2)_n$ flexible spacer group, n = 2, 6 or 12

Polymer 3
Backbone-COO-C_6H_4-CH=N-C_6H_4-O-C_2H_5

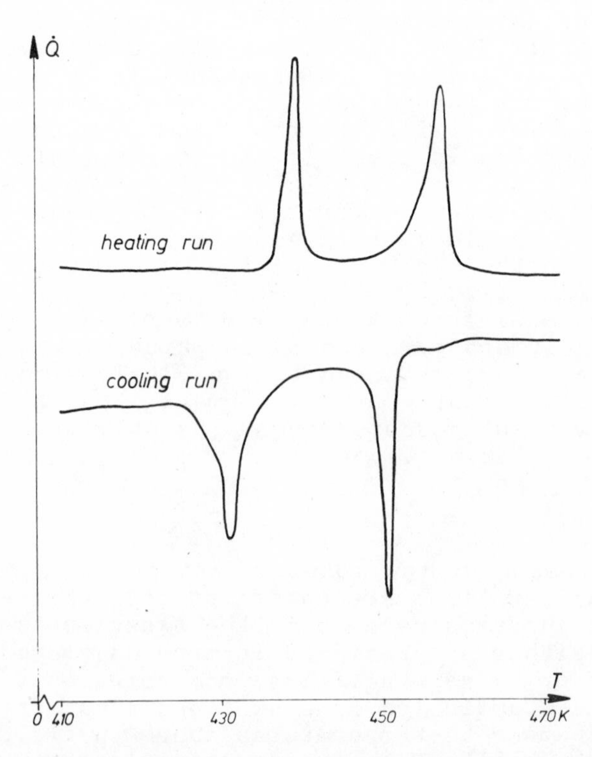

Figure 1. DSC traces of a polymer displaying a smectic and a nematic phase

with a Kratky small angle diffraction unit and with a wide angle diffractometer. The scattered intensity was registered stepwise using a scintillation counter combined with an impuls height discriminator. Diagrams of oriented samples- due to induced textures, were obtained by means of a flat film camera. The position and the width of the reflections were calibrated by comparison with the diagrams of quartz. Additional information about the structures and the stability region of the structures were obtained by thermal analysis (DSC) and by means of a polarizing microscope.

Thermal Properties

Polymers with flexible spacer groups. Polymers obtained by precipitation or by casting films from solutions were found to be amorphous as shown by X-ray and microscopical observations. The DSC data revealed that on heating up the samples for the first time a stepwise increase of the specific heat occurred which could be attributed to a glass transition. Thus at room temperature the "as received" samples were in the isotropic glassy state, apparently no liquid crystalline solutions exist.

Just above the glass transition temperature an exothermic peak was observed, indicating that an ordered structure was formed. Subsequent X-ray studies showed that either nematic or smectic structures were formed, depending on the nature of the substituent R. On further heating one or two sharp melting peaks were observed (Figure 1), the polymers were transformed into the isotropic fluid state, as shown by X-ray data. On cooling the samples the transformation into liquid crystalline states occurred as shown by exothermic peaks (Figure 1). At lower temperatures a stepwise decrease of the specific heat was found, thus at room temperatures the samples are nematic or smectic glasses. The melting and crystallization ranges were very narrow of the order of 1 or 2 deg. C. The degree of super-cooling was very low, at a heating and cooling rate of 4 deg. per minute the difference in melting and crystallization temperatures amounted to about 2 deg. C. In the case of the nematic transition the heats of fusion were of the order of 2 Joules/g whereas in the case of the smectic transitions the heats of fusion were of the order of 10 Joules/g. The properties discussed above are very unusual for polymeric systems but quite usual for low molecular weight liquid crystalline systems.

The melting points of the polymers studied were in the range of 100 to 130 deg. C for polymers 1 and around 200 deg. C for polymers 2.

At the glass transition a distinct stepwise change of the specific heat was only observed for nematic phases whereas for the case of the smectic phase only a change in the slope of the specific heat temperature curve was found. Additional techniques such as dilatometry have to be used in order to prove that actually a glass transition occurs. This result indicates that since the smectic structure is closer to that of a crystal, the motions which take place above the glass transition will be restricted. Thus the increase in the specific heat will be small. The types of motions which freeze in at the glass transition will be discussed below.

Isotropic glasses are characterized by relaxation processes, which are directly related to the non-equilibrium nature of the glassy state. Similar processes should also occur in liquid crystalline glasses. Enthalpy relaxation was observed for samples, which had been annealed below the glass transition temperature for several hours, proving the non-equilibrium nature of smectic and nematic glasses.

Polymers without flexible spacer groups. The DSC curves of the polymer 3 indicated the existence of a liquid crystalline glassy state at room temperature. The polymer was found to be smectic. Two melting peaks were observed in the temperature range between 300 and 310 deg. C. These peaks are not as easily observed as in the case of the polymers discussed above, since the decomposition takes place in the same temperature range. The occurrence of exothermic peaks on cooling nevertheless indicates that reversible melting and crystallization processes take place.

X-Ray Investigations

Polymers with flexible spacer groups. Smectic or nematic polymers were obtained depending on the nature of the substituent R in the case of polymers 1 (Table 2). It has to be pointed out that the structures at temperatures above the glass transition temperatures are equilibrium structures, since thermodynamically stable liquid crystalline phases exist. This is in contrast to most systems studied until now where only liquid crystalline structures were obtained.

The nematic phases were characterized by one broad halo in the wide angle scattering region, which could

Table II

Thermal properties of selected examples of polymers 1 (T_S smectic transition, T_N nematic transition, H heat of fusion)

Nematic polymers			
n=2	R=OCH_3	T_N=121 °C	H= 2.3 Joules/g
n=6	R=OCH_3	T_N=105°C	H= 2.1 Joules/g
Smectic polymers			
n=2	R=OC_3H_7	T_S=129 °C	H= 9.2 Joules /g
n=2	R=OC_6H13	T_S=140°C	H=11.3 Joules /g
n=6	R=OC_6H_{13}	T_S=115°C	H=15.5 Joules /g
Diphenyl derivatives			
n=2	R=C_6H_4-OCH_3	T_N=174°C	H= 3.1 Joules /g
n=2	R=C_6H_5	T_S=124°C	H= 1.0 Joules /g
		T_N=187°C	H= .8 Joules /g
n=6	R=C_6H_5	T_S=160°C	H= .8 Joules /g
		T_N=181°C	H= 2.3 Joules /g

be attributed to the short range positional order of the side groups characteristic for the nematic order. The molecular distances derived from these data corresponded to those of parallel side groups. No reflections were observed which could be attributed to a positional order in the direction of the long axes of the side groups, thus no cybotactic clustering took place [8]. The scattering diagram is very similar to that of amorphous polymers. One consequently has to use additional techniques such as polarizing microscope techniques in order to verify a nematic structure. The nematic structure is characterized by a local orientation of the side groups. In the absence of a specific texture no macroscopic orientation will results since the preferred direction changes within the sample. Slightly orientated samples were obtained by drawing, highly oriented samples were obtained if specific textures were induced. This will be discussed later. No information was gained about the conformation of the chain. It can be assumed however, that the conformation is very close to that of the isotropic fluid state due to the flexible spacer groups.

The smectic structure was characterized by one or more (higher order) reflections in the small angle scattering region, which were sharp, and by a broad halo in the wide angle region. The small angle reflections could be attributed to smectic layers, the wide angle halo to a two-dimensional fluid state within the smectic layers. Thus the structure is characteristic of a smectic A or C modification. In order to obtain more information about the smectic structures in polymers extended studies including studies on oriented systems were performed on a specific polymer displaying a smectic structure; a polymer without a flexible spacer group was chosen. This will be discussed below.

In the case of the polymers 2 smectic structures were observed for the homopolymers, that is for polymers in which the flexible spacer group was constant.

In addition to the homopolymers also copolymers were studied, in which the comonomers were characterized by spacer groups of different length. The combinations studied were n = 2/6; n = 6/12 and n = 2/12. Smectic polymers were observed by copolymerization of monomers with n = 2/6 and n = 6/12, whereas cholesteric polymers were obtained by copolymerization of monomers with n = 2/12, if the composition was approximately 1 : 1. Apparently large differences in spacer length, which lead to shifts of the mesogenic cholesteryl group relative to each other are required for the destruction of the smectic order. The X-ray pattern of the homopolymers and the copolymers 2/6 and 6/12 are characterized by sharp reflections in the small angle region, whereas the pattern of the copolymer 2/12 is characterized by increasingly broader reflections if the composition 1 : 1 is approached. At this composition only a broad small angle halo is observed. The existence of a cholesteric structure is established by means of polarizing microscope techniques and by the observation of selective reflection. The side groups are parallel to a prefered direction in these polymers on a local scale, the prefered direction changes in the sample in such a way that a helical structure results.

Polymers without flexible spacer groups. The tendency to form smectic structures is increased if the spacer group is missing. This is probably caused by the strong coupling between the position of neighbouring mesogenic side groups due to the chain backbone. Oriented samples-used for structural analysis, were obtained by performing the polymerization in the nematic phase of the monomer under the influence of a magnetic field. It should be pointed out that the final structure of the polymer did not depend on the method by which the polymer was synthesized. In all cases the diffraction pattern was characterized by four sharp reflections in the small angle region and one broad reflection in the wide angle region. The small angle reflections corresponded to a lattice dimension of 20.3 Å, which was related to the dimension of the smectic layers.

The small angle reflections were analyzed with respect to the dependence of their width on the order of the reflections. From this analysis information on structural defects of the layer arrangement can be obtained. According to the theory of paracrystalline distortions [9] the integral width δB of the reflections can be related to the size L of smectic aggregats (crystal size in the direction of the layer normal) and to the relative fluctuations $\delta l/l = g$ of the lattice dimension (smectic layer dimension) l by the expression:

$$\delta B^2 = 1 / L^2 + (\pi g h)^4 / l^2$$

where h is the order of the reflection. It was found that the dependence of the width of the reflections on the order could be represented by this equation.

From the intercept and the slope of the line a value of 1580 Å was obtained for L and a value of 1.4 % for g. Thus up to 80 layers are within one smectic aggregat, the relative fluctuation of the layer dimension is very small.

Additional information can be obtained from oriented samples, they were characterized by small angle reflections on the meridion and a broad reflection in the wide angle region on the equator. The reflections were elongated along Debye rings, due to misorientations. The fiber diagram revealed that the direction of the one dimensional long range order, which was identical with the direction of the fiber axis, was perpendicular to the direction, in which only a short range order existed. Thus the smectic structure of the polymer is characterized by assemblies of smectic layers in the direction of the layer normal and a short range order within the smectic layers. In these layers the side groups, containing the mesogenic group, are parallel to each other and parallel to the layer normal as in the case of the smectic A modification. If one compares the molecular dimensions of the chain with the values obtained from the scattering experiment one has to conclude that the polymer chains are arranged in a staggered way. Neighbouring polymer chains must be shifted in a regular way in the direction of the long axes of the side groups. It is possible that the CH_3 groups of the methacrylic backbone disrupt the close packing of the side groups since in the case of an acrylic backbone no shift of the neighbouring side chains was observed.

In the smectic polymers the main chain must be confined to parallel planes within the smectic layers since the observed periodic structure in the direction of the layer normal can only be obtained by this arrangement. The X-ray data do not yield information on the actual two dimensional conformation of the chain on these smectic planes.

On heating up the sample the X-ray diagram remained unchanged up to the glass transition temperature. Above this temperature the small angle reflections remain unchanged whereas the width and the position of the wide angle halo change in a way characteristic of increased molecular distances and a broader distribution of the molecular distances. Thus the motions which set in at the glass transition occur only within the smectic layers in a direction parallel to the layer surface. The smectic layer acts as a two-dimensional fluid.

Liquid Crystalline Textures

Liquid crystalline textures in polymeric systems can be obtained in two different ways. One way consists in inducing a specific texture in a liquid crystalline monomer phase. Often the forces acting on the sample will be sufficiently large to stabilize the texture during the polymerization. A texture may be obtained in this way even in polymers which do not display liquid crystalline phases. In the case of polymer 3 the polymerization in the nematic monomer phase under the influence of a magnetic field led to a polymer sample with high orientational order on a macroscopic scale.

A second way to obtain liquid crystalline textures consists in heating up polymer samples, which display liquid crystalline phases, between glass slides or under the influence of external forces such as electric or magnetic fields. Textures will then form in a way similar to the case of low molecular weight liquid crystals. This was observed for the case of the polymers 1. On cooling down the textures can be frozen in since a glass transition occurs. Thus it becomes possible to keep a specific texture with interesting optical properties permanently.

Literature Cited

1) Blumstein, A., "Liquid Crystalline Order in Polymers", Academic Press - in press
2) Shibaev, V.P., Vysokomol. Soedin (1977) A9, 923
3) Blumstein, A., Macromolecules (1977) 10, 872
4) Flory, P.J., Proc.Roy.Soc.Lond. (1956) A234, 73
5) Jackson, W.J. and Kuhfuss, H.F., J. Pol. Sci.-Polymer Chem. Ed. (1976) 14, 2043
6) Roviello, A. and Sirigu, A., Polymer Letters Ed. (1975) 13, 455
7) Finkelmann, H., Ringsdorf, H. and Wendorff, J.H. - this issue
8) De Vries, A., Mol.Cryst.Liquid Cryst. (1970) 10, 31, 219
9) Hosemann, R. and Bagchi, S.N., "Direct Analysis of Diffraction by Matter", North Holland Publ. Comp., Amsterdam 1962

RECEIVED December 8, 1977.

3

Enantiotropic (Liquid Crystalline) Polymers: Synthesis and Models

H. FINKELMANN, H. RINGSDORF, W. SIOL, and J. H. WENDORFF

Institut für Organische Chemie, Universität Mainz and Deutsches Kunststoff Institut, Darmstadt, West Germany

In the past few years studies on liquid crystalline polymers have attracted increasing interest because of their theoretical and technological aspects (1). In order for the liquid crystalline phase to form, the molecules must have an extended rigid molecular shape and a large anisotropic polarizability (2). If a polymer has to behave like a conventional liquid crystal, it is best to use a mesogenic molecule as a monomer component. By substituting a vinyl group, the polymerization leads to an alkyl polymer chain, where the mesogenic moieties are fixed as side chains. In the liquid state, the polymer main chain is flexible, and only the rigid mesogenic side chain moieties are required to form the mesophase. They have to build the liquid crystalline order independently from the conformation of the connecting main chain. Consequently the theoretical considerations used in the case of conventional liquid crystals have to be valid for this type of polymer.

In the past few years attempts to synthesize liquid crystalline polymers have started with a monomer in a liquid crystalline state. Three different types of polymer systems have been obtained:

(1) Amorphous polymers -- which exhibit no liquid crystalline properties.

(2) Polymers, with fixed liquid crystalline structure -- if the polymerization temperature was below the glass transition temperature (T_G) of the polymer. These polymers exhibit a liquid crystalline structure in the solid glassy phase because the structure of the ordered monomer phase was "frozen in" by polymerization. By heating the polymer, these frozen structures were irreversibly lost above the T_G (4,5). An additional effect was observed; starting from a nematic or cholesteric monomer in most cases, the resulting polymer structure was smectic (6,7). The decrease of mobility of the mesogenic side chains because of their

0-8412-0419-5/78/47-074-022$05.00/0

fixation to the polymer main chain implies an additional long range order of the centers of gravity of the mesogenic groups.

(3) Liquid crystalline (LC) polymers -- with a thermodynamically stable liquid crystalline phase above the T_G. These were obtained only in a few cases (8,9).

Until now there was no obvious correlation found between the monomer structure and the resulting polymer phase. No theoretical structural conditions were described which would result in a liquid crystalline polymer with a definite ordered phase, e.g., with a nematic, a smectic, or a cholesteric phase as in conventional liquid crystals. Although previous examples have established (8,9) the existence of enantiotropic liquid crystalline side chain polymers, additional considerations are in order for a systematic synthesis of such polymers.

Model Considerations

In conventional LC phases the motion of the molecule is restricted only by the anisotropic interactions with its neighbors. This leads to the formation of the orientational long range order and, in the case of smectic phases, to an additional lamellar structure. However, completely different conditions normally exist in a liquid crystalline polymer. Figure 1 schematically shows the typical structure for a LC polymer as described so far. The mesogenic groups are directly fixed to the main chain. Thus for the liquid crystalline polymer state above the T_G two conditions have to be considered.

(1) The polymer main chain has to be in the liquid state, causing a tendency toward a statistical chain conformation.

(2) The mesogenic side chains have to be in the liquid crystalline order. The anisotropic interactions induce an anisotropic orientation.

These two tendencies conflict, and steric hindrance in the system determines which tendency will dominate. The steric hindrance results from the direct linkage of the main chain to the side chain. A direct coupling of motions of the main chain to the rigid mesogenic side chains can be assumed. Therefore, an amorphous liquid polymer is obtained if the anisotropic orientation of the side chain is hindered or disturbed by the main chain. On the other hand, a solid polymer with a LC structure is obtained if the anisotropic-ordered rigid side chains hinder the normal motions of the main chain and thus tend to restrict the main chain motions.

The critical point seems to be the direct linkage between the main and side chains. Thus one can assume that to obtain a polymer in its liquid crystalline state, the direct coupling has to be avoided. If this linkage were flexible, steric hindrance would be alleviated, and the motions of the side and main chains would be decoupled (Figure 2). Under these conditions a statistical main chain conformation as well as an anisotropic orientation of the side chain would occur, resulting in an enantiotropic LC polymer. The flexible linkage (schematically represented in Figure 2 by ∼∼) can be realized by an alkyl- or alkyloxy chain and will be called "flexible spacer group." If the model with the decoupled main chain and side chain with the flexible spacer group is valid, the properties of the resulting liquid crystalline polymer should be determined only by the mesogenic moieties similar to conventional LC.

The model can be applied by homo- and copolymerization (Figure 3). Nematic as well as smectic polymers may be expected with homopolymerization. With regard to a homologous series of a mesogenic components with varying para substituents, nematic phases should be formed preferentially by using short para substituents. Smectic phases should be found preferentially if the para substituent is sufficiently long. This analogy to low molecular liquid crystals is expected if the main and side chains are decoupled. The knowledge of conventional liquid crystals would then be applicable to these types of polymers. With copolymerization basically the same principles can be expected as for the homopolymers. Using different monomers, a wide variation of the temperature region of the mesogenic phase may occur. An additional effect has to be expected with copolymerization of monomers with spacers of different lengths. The resulting copolymers should exhibit even less order than the corresponding spacer-group-containing homopolymers. It should be possible to favor an arrangement of mesogenic side groups in which the centers of gravity are distributed statistically and thus avoid the smectic order. These models might be of interest, especially for the formation of nematic polymers.

Results and Discussion

Homopolymers. Phenylesters of benzoic acid were chosen as mesogenic groups for the synthesis of suitable monomers. The acryloyl- and the methacryloyl moieties were used as polymerizable groups. The flexible spacer was an alkyl chain or an alkyloxy chain of varying length. Thus a homologous series of p-Ω-(2-methylpropenoyloxyalkyoxy) benzoic acid -p' substituted phenyl esters 1 was prepared. The monomers were synthesized by the standard methods as follows:

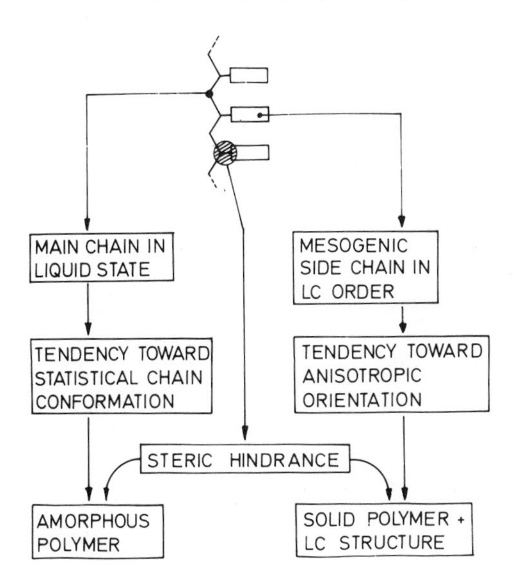

Figure 1. Schematic of formation of liquid crystalline order in polymers with mesogenic side groups attached directly to the main chain

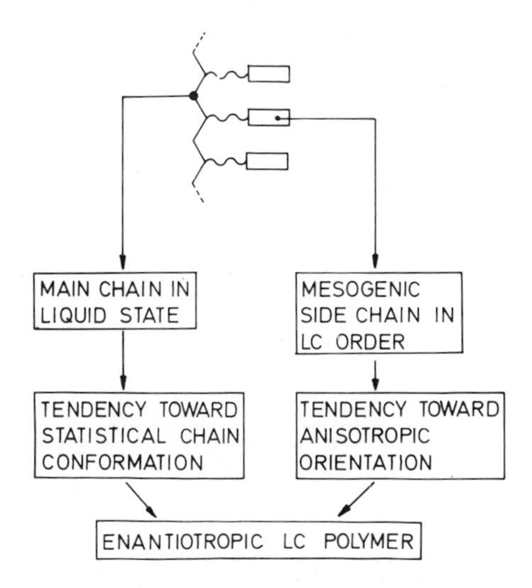

Figure 2. Schematic of liquid crystalline order formation in polymers with mesogenic side groups decoupled from the main chain by a flexible spacer

1. HOMOPOLYMERIZATION

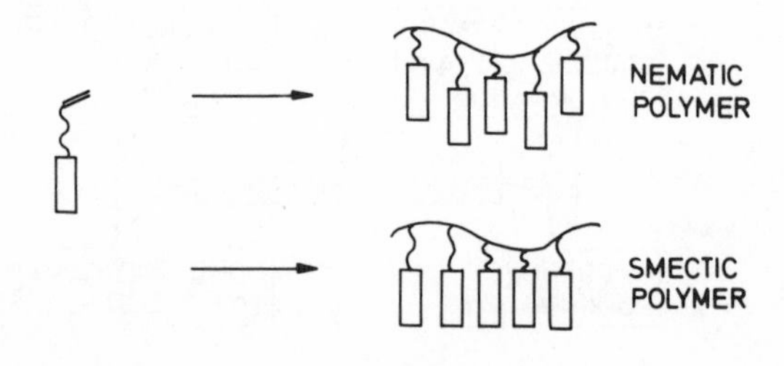

2. COPOLYMERIZATION

Figure 3. Schematic of nematic and smectic order in (a) copolymers and (b) homopolymers

$$HO-(CH_2)_{\underline{n}}-Cl + HO-C_6H_4-COOH \xrightarrow{NaOH} HO-(CH_2)_{\underline{n}}-O-C_6H_4-COOH$$

$\underline{n}$ = 2, 3, 6

$$\xrightarrow[H^+]{CH_2=C(CH_3)-CO-OH} CH_2=C(CH_3)-CO-O-(CH_2)_{\underline{n}}-O-C_6H_4-COOH \xrightarrow[2.\ HO-C_6H_4-R]{1.\ SOCl_2}$$

$$CH_2=C(CH_3)-CO-O-(CH_2)_{\underline{n}}-O-C_6H_4-COO-C_6H_4-R \qquad \underline{1}$$

Flexible spacer — Mesogenic group

The properties of some monomers are listed in Table I. Only the diphenyl derivatives exhibit enantiotropic liquid crystalline properties whereas the phenylesters melt to isotropic fluids. A monotropic liquid crystalline phase is obtainable if long spacers and para substituents are used. Because of the high tendency to crystallize, the monotropic phase transitions of the monomers were not determined.

All monomers are radically polymerizable. In accordance with the predictions derived from the model, these polymers exhibit enantiotropic liquid crystalline properties. Even the polymers prepared from isotropic monomers are liquid crystalline. Table II lists the phase transition and LC phases which were determined by DSC, x-ray investigations(10), and polarization microscopy.

If one compares the polymers where R is an alkyloxy group, nematic as well as smectic phases are observed. If R is a short substituent ($-OCH_3$), the polymer is nematic, whereas if R is a longer substituent, the polymer is smectic. The diphenyl derivatives exhibit same characteristic. In addition the diphenyl derivatives without a para substituent have a low temperature smectic phase as well as a nematic phase.

These results show that the side chain polymers fit the model well. Furthermore they confirm the expected analogy of these spacer-group-containing polymers with conventional liquid crystals that nematic and smectic phases can be realized, depending on the para substituents of the mesogenic moieties.

Copolymers. The second part of our model is concerned with the copolymerization of monomers with different spacers. In this case a statistical distribution of the centers of gravity of the mesogenic chain might be preferred even more. This prediction was tested with cholesteryl derivatives:

Table I. Phase Transitions of Monomers 1

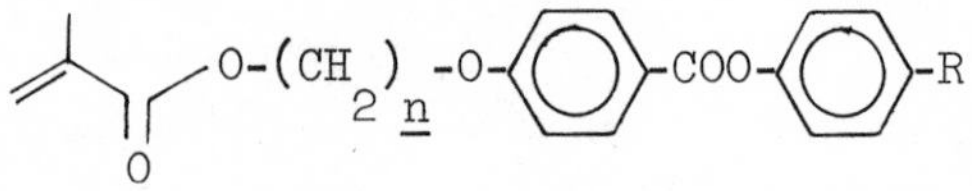

$CH_2=C(CH_3)-C(=O)-O-(CH_2)_n-O-C_6H_4-COO-C_6H_4-R$

1. Alkylethers

n	R	Phase Transition (°C)
2	OCH_3	k 69 i
2	OC_3H_7	k 67 i
2	OC_6H_{13}	k 59 i
6	OCH_3	k 47 i
6	OC_6H_{13}	k 47 n 53 i

2. Diphenyl derivatives

n	R	Phase Transition (°C)
2	$C_6H_4OCH_3$	k 108 n 211 i
3	$C_6H_4OC_2H_5$	k 123 n 202 i
3	C_6H_5	k 105 i
6	C_6H_5	k 64 sm 68 n 92 i

Table II. Liquid Crystalline Phase Transition of Synthesized Polymers

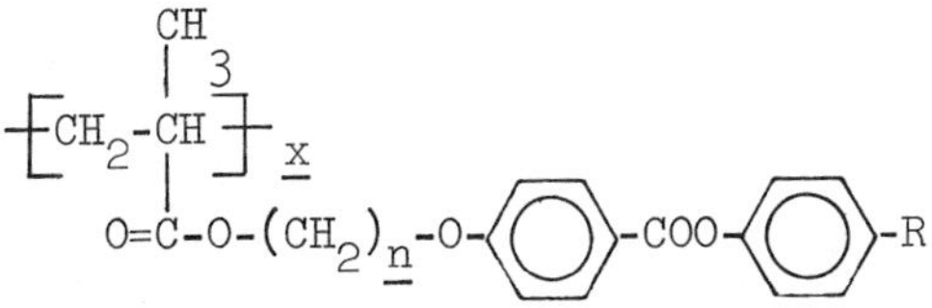

$$-\!\!\left[CH_2-\underset{\underset{O=C-O-(CH_2)_n-O-C_6H_4-COO-C_6H_4-R}{|}}{\overset{\overset{CH_3}{|}}{CH}}\right]\!\!-_x$$

1. Alkylethers

a) Nematic polymers

n	R	Phase Transition
2	OCH_3	g 101 n 121 i
6	OCH_3	g 95 n 105 i

b) Smectic polymers

n	R	Phase Transition
2	OC_3H_7	g 120 sm 129 i
2	OC_6H_{13}	g -[a]) sm 140 i
6	OC_6H_{13}	g 60 sm 115 i

2. Diphenyl derivatives

n	R	Phase Transition
2	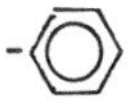 $-C_6H_4-OCH_3$	g -[a]) 177 i
3	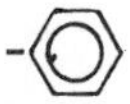 $-C_6H_4-OC_2H_5$	g 120 sm 300 i
3	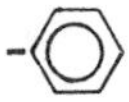 $-C_6H_4-$	sm 170 n 197 i
6	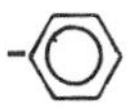 $-C_6H_4-$	g 132 sm 164 n 184 i

[a] Not yet determined.

O-⟨◯⟩-$(CH_2)_n$-C(=O)-O-chol 2

n = 2, 6, 12

in which the spacer length n is varied. All synthesized monomers exhibit cholesteric mesophases instead of nematic phases because of the chirality of cholesterol (Table III).

Table III. Phase Transitions of Cholesteryl Derivatives 2

O-⟨◯⟩-$(CH_2)_n$-C(=O)-O-chol 2

n	Phase Transition (°C)
2	k 139 ch 159 i
6	k 90.5 ch 112 i
12	k 55 ch 81 i

The synthesized monomers are radically polymerizable, and their polymers and copolymers were purified by reprecipitation. Table IV summarizes the results of polymerization experiments. In all cases enantiotropic liquid crystalline polymers were obtained. The homopolymers exhibit smectic phases. Contrary to our expectations smectic polymer phases are obtained also by copolymerization of the monomers in which the spacer length is n_1 = 2, n_2 = 6 and n_1 = 6, n_2 = 12. A similar result was published recently by Shibaev et al. (7), who, based on investigations of comb-like polymers, synthesized polymer cholesteryl derivatives with enantiotropic smectic phases.

The expected preference of a statistical distribution of the mesogenic side chain has so far been only realized by the copolymerization in which n_1 = 2, n_2 = 12 and only if the monomer ratio is 1:1. Other compositions also led to smectic phases. The best proof for its cholesteric structure is its selective reflection of left-circular polarized light in the visible, which indicates a left-handed cholesteric helix of the polymer. The structure has been established also by x-ray investigations (8). This copolymer is the first enantiotropic cholesteric polymer.

This result confirms the model and shows that a statistical distribution of centers of gravity of the mesogenic side chains is possible if spacer groups of sufficiently different lengths are used.

Table IV. Phase Transitions of Homo- and Copolymers of Cholesteryl Derivatives 2 [a]

Homopolymers

n	T_{Kl} (°C)[b]	$\Delta H_{T_{Kl}}$ (J/g)[b]	Polymer Phase[c]
2	-[d]	-[d]	smectic
6	182	6.7	smectic
12	168	4.2	smectic

Copolymers

n_1	n_2	M_1	M_2	T_{Kl}[b]	$\Delta H_{T_{Kl}}$ (J/g)[b]	Polymer Phase[c]
2	6	50	50	-[d]	-[d]	smectic
6	12	50	50	185	2.76	smectic
2	12	51	49	209	1.17	cholesteric
2	12	56	44	212	1.25	smectic
2	12	34	66	-	-	smectic
2	12	42	58	-	-	smectic

[a] n_1, n_2 = spacer length of components M_1, M_2.

[b] DSC measurements.

[c] X-ray investigations

[d] Decomposition before clearing.

Bulk Polymerization. The monomers were also polymerized in bulk. To follow the process, a polarization microscope was used, which made it possible to observe phase separations or phase transitions during polymerization.

(1) A phase separation during polymerization occurred when a nematic monomer was polymerized to a smectic polymer, because the polymer was insoluble in the monomer.

(2) No phase separation was observed when an isotropic monomer was polymerized to a nematic or smectic polymer. The nematic or smectic polymers are soluble in the isotropic monomer. When a certain conversion is obtained, a phase transition from isotropic to nematic or smectic phase takes place. This phase transition, caused by variation of the mole fraction of the monomer and the polymer, is comparable with a phase transition caused by varying the temperature at constant composition.

(3) A homogenous polymerization was achieved by the polymerization of a nematic monomer to a nematic polymer where the temperature region of the liquid crystalline monomer phase overlapped the temperature region of the liquid crystalline polymer phase (see Table II). This polymerization process is influenced either by a precipitation or by a phase transition.

A detailed physical investigation of the polymers described here is discussed in Chapter 2.

Literature Cited

1. Shibaev, V.P., Vysokomol. Soedin (1977) A9, 923.
2. Gray, G.W., Winsor, P.A., "Liquid Crystals and Plastic Crystals," Ellis Horwood, Chichester, 1974.
3. Strzelecki, L., Liebert, L., Bull. Soc. Chim. (1973) N2, 605.
4. Clough, S.B., Blumstein, A., Hsu, E.C., Macromolecules (1976) 9, 123.
5. Bouligand, Y., Cladis, P.E., Liebert, L., Strzelecki, L., Mol. Cryst., Liq. Cryst. (1974) 25, 233.
6. Lorkorwski, H.J., Reuther, F., Plaste Kautsch. (1976) 2, 81.
7. Hsu, E.C., Blumstein, A., Clough, S.B., J. Polym. Sci., Polym. Lett. Ed. (1977) 15, 545.
8. Perplies, E., Ringsdorf, H., Wendorff, J.H., Ber. Bunsenges. Phys. Chem. (1974) 9, 921.
9. Shibaev, V.P., Freidzon, J.S., Platé, N.A., Dokl. Akad. Nauk, SSSR (1976) 227, 1412.

RECEIVED March 2, 1978.

4

Thermotropic Cholesterol-Containing Liquid Crystalline Polymers

V. P. SHIBAEV, N. A. PLATÉ, and Y. S. FREIDZON

Polymer Chemistry Department, Faculty of Chemistry,
Moscow State University, Moscow, USSR

In recent years, investigators working in the field of physics and chemistry of high-molecular compounds have been paying a lot of attention to the problem of creating liquid crystalline systems (1-14). The great interest displayed in the study of properties of such systems can most probably be accounted for by two main factors: firstly, by the advances in studies into the structure, properties and practical use of low-molecular liquid crystals in physics, technology and medicine, and, secondly, by the studies of the nature and salient features of the liquid crystalline state in polymers as a specific state of macromolecular substances.

Advances in this field of research are associated with the development of methods for producing polymeric liquid-crystalline systems and controlling the processes of structure formation in polymers.

Unfortunately, at present, the relevant literature not only lacks classification of experimental data on polymeric liquid crystals, but also criteria determining the existence of the liquid crystalline state in polymers are nowhere to be found. More often than not, certain authors refer to polymer systems under study or specific states of these systems as liquid-crystal ones without sufficient grounds.

The recently published reviews by Papkov on lyotropic liquid crystalline polymer systems (12) and by Shibaev and Plate on the liquid crystalline states in polymers (13) should be regarded as the first attempts to systematize the great body of available experimental data with a view to elaborating adequate techniques of producing liquid crystalline polymer systems.

In this paper, which deals exclusively with thermotropic liquid crystalline polymers, we shall apply the term "liquid-crystalline" to the thermodynamical-

0-8412-0419-5/78/47-074-033$06.00/0

ly stable phase state of polymers or polymer systems characterized by a spontaneously occurring (independently from their state of aggregation) anisotropy of properties (in particular, optical anisotropy) in the absence of a three dimensional crystal lattice. It should be pointed out that along with this definition which describes the liquid crystalline state as a phase state, use is often made of the term "liquid crystalline structure" which is indicative only of a certain orientation ordering in a system. Despite the narrower meaning of the latter term, the notion of a liquid crystalline structure (or ordering) is widely used in the literature on structural polymer studies, therefore, in some instances, we shall use this term as well.

At present, four types of polymer systems can be distinguished to which, in our opinion, the term of liquid crystalline (mesomorphous) state is applicable:

(I) Melts of crystalline polymers and amorphous polymers characterized by an orientation related to liquid crystalline ordering;

(2) Lyotropic liquid crystalline systems;

(3) Mesomorphous structures of block polymers in gels;

(4) Polymers with side anisodiametric groups modeling the molecular structure of low-molecular liquid crystals.

In this paper, we shall dwell on approaches to creating liquid crystalline polymers of the latter type. As far as the first three systems are concerned, their description can be found in Refs. (12-13) and the cited references.

Polymers with side anisodiametric groups modeling the molecular structure of low-molecular liquid crystals can be obtained either through synthesis of monomers with liquid crystalline (mesogenic) groups with subsequent polymerization or through chemical attachment of molecules of low-molecular liquid crystalline compounds to a polymer chain by way of polymer-analogous transformations. As can be inferred from the bulk of works dealing with this problem, at present, the first of these two methods is used in most cases. Ref. (13) reviews the basic types of the so far synthesized polymers with mesogenic groups along with a detailed description of the structure of these compounds. Analysis of these data suggests that the presence of mesogenic groups directly linked with the main chain (Figure 1), as a rule, does not result in such polymers manifesting liquid crystalline properties. If some monomers do in fact exhibit liquid crystal properties, the polymers obtained on their basis are rigid substances featuring high softening points without displaying truly liquid crystalline properties in accordance with our definition. Only some works (9,10,15-18) provide evidence that the synthesized polymers possess the property of birefringence, however, no detailed structural and thermodynamic characteristics of these polymers are given.

In our opinion, such a macromolecular structure in which the mesogenic groups are directly linked with the main polymer chain renders difficult the packing typical for low-molecular liquid crystals. To obtain an liquid crystalline structure, a certain lability of branches is required, which would, despite the presence of the polymer chain, ensure a particular ordering in the arrangement of mesogenic groups. To overcome the steric hindrances occurring in the packing of branches, one should space the mesogenic groups a certain distance apart from the main chain, or somehow enhance its flexibility.

In considering various approaches to creating thermotropic cholesterol-containing polymers (11,19-20), we proceeded from the derived notions on the relationship between the structure and properties of comb-like polymers with long aliphatic branches in each monomer unit (21). The independent behaviour of the side chains of comb-like polymers, manifesting itself in their ability to form layered structures and even to crystallize, regardless of the main chain's configuration (21), opens up possibilities to obtain liquid crystalline comb-like polymers. Indeed, since the chemical bonding of asymmetric side pendants in such polymers takes place only through the end groups of branches, the latter may be regarded as a structurally organized arrangement of long-chain molecules, built around the main chain, which seems to be one of the prerequisites for a liquid crystalline structure. It could, therefore, be assumed that if one would add, to the end groups of comb-type polymer branches, groups capable of forming the liquid crystalline phase, i.e. space them a certain distance apart from the main chain, the steric hindrances imposed by the main chain on the packing of branches would be much less significant.

As mesogenic groups we have selected cholesterol derivatives having, in the case of low-molecular liquid crystals, most extensive application. To this end, we have synthesized cholesterol esters of N-methacryloyl-ω-aminocarbonic acids (ChMAA-n) having different lengths of the aliphatic radical (index n corresponds to the number of methylene groups in the alkyl radical linking cholesterol to the main chain), as follows:

$$H_2C=C(CH_3)-C(=O)-Cl + H_2N-(CH_2)_n-C(=O)OH \xrightarrow{-HCl} H_2C=C(CH_3)-C(=O)-HN-(CH_2)_n-COOH \quad (I)$$

$$H_2C=C(CH_3)-C(=O)-HN-(CH_2)_n-COOH + \text{HO-cholesteryl}$$

$$H_2C=C(CH_3)-C(=O)-HN-(CH_2)_n-COO-\text{cholesteryl} \quad (II)$$

where n = 2,5,6,8,10,11.

The first stage yielded N-methacryloyl-ω-aminocarbonic acids (MAA-n) whose subsequent reaction with cholesterol yielded monomers of ChMAA-n. From these monomers, homopolymers (PChMAA-n) and copolymers with n-alkylacrylates (A-m) and n-alkylmethacrylates (MA-m) were obtained by radical polymerization (index m corresponds to the number of carbons in the n-alkyl radical). In addition, in order to elucidate some of the problems relating to the structure of liquid crystalline polymer compounds, monomers were specially synthesized and polymers were obtained containing methyl (PMMAA), benzyl (PBMAA) and hexadecyl (PHMMA) groups in the side chain instead of cholesterol:

$$H_2C=C(CH_3)-CONH-(CH_2)_n-COO-R,$$

where R = $-CH_3$ for n = 2,5,6,8,10,11;
(PMMAA-n)
$-CH_2-C_6H_5$ for n = II (PBMAA-II);
$-(CH_2)_{15}-CH_3$ for n = II (PHMAA-II).

It should be noted that Ref. (22) describes synthesis of cholesterol esters of poly-N-acryloyl-ω-aminocarbonic acids through addition of cholesterol to the macromolecules of poly-N-acryloyl-ω-aminocarbonic acids. However, the resulting polymers contained about 10 mol. % of carboxyls and did not exhibit any liquid crystalline properties. An interesting example of synthesis of comb-like polycholesteryl-II-methacryloyloxyundecanoate has been described by Imoto et al. (23), however, apart from the study of the phase state of the monomer, no data are provided on the liquid crystalline structure of the polymer. This work is, therefore, aimed at:

(I) developing methods for obtaining cholesterol-containing monomers, as well as polymers with different length of the side chain and frequency of occurrence of cholesterol groups;

(2) studying the physicochemical behaviour of monomers of ChMAA-n and cholesterol-containing polymers in the solid phase, as well as defining the conditions under which the polymers and copolymers under consideration acquire the liquid crystalline structure.

Experimental

Synthesis of Monomers N-methacryloyl-ω-aminocarbonic acids (MAA-n) were obtained as follows

Added to a solution of 0.08 M of ω-aminocarbonic acid in a 0.6 n solution of NaOH (600 ml), with stirring, were 0.25 M of chloride of methacrylic acid. After 3 hours of stirring, the reaction mixture was acidified with HCl up to pH = 3. The precipitate was filtered, dissolved in chloroform and washed, successively, with a solution of Na_2CO_3, a weak solution of HCl (pH = 2-3), and water. After the solution had been dried over $MgSO_4$, it was evaporated, and the monomer was recrystallized from the acetone-petroleum either mixture. The yield was 70%. The melting points of MAA-n are given in Table I.

Table I. Melting points of N-methacryloyl-ω-aminocarbonic acids (MAA-n)

n	2	5	6	8	10	11
T_m, °C (±1°)	67	53	43	48	68	72

Cholesterol esters of N-methacryloyl-ω-aminocarbonic acids (ChMAA-n) were obtained as follows (20):

0.02 M of MAA-n, 0.02 M of cholesterol and 0.05 g of p-toluene sulfonic acid were dissolved in 200 ml of absolute benzene. The mixture was heated to the boiling point and the azeotropic mixture of benzene with the resulting water was distilled off. 5 hours later, the reaction mixture was cooled down to room temperature, and the monomer was precipitated with methanol. The monomer was freed of cholesterol on a column with Al_2O_3. The yield was 40%. The results of the elemental chemical analysis of the monomers of ChMAA-n are listed in Table II.

Table II. Elemental analyses of ChMAA-n monomers

n	C, %		H, %		N, %	
	found	calculated	found	calculated	found	calculated
2	77.46	77.66	10.61	10.54	2.52	2.66
5	78.48	78.25	10.49	10.82	2.55	2.47
6	78.42	78.43	11.10	10.91	2.36	2.41
8	78.85	78.76	10.58	11.07	2.41	2.30
10	79.40	79.06	10.95	11.21	2.12	2.19
11	79.52	79.20	11.51	11.28	2.12	2.15

Cholesterylmethacrylate (ChMA) were obtained by way of interaction of chloride of methacrylic acid with cholesterol in absolute ether. The melting point of the end product was 109°C, which coincides with that of ChMA obtained in (24).

Hexadecyl and benzyl ethers of MAA-II (HMAA-II and BMAA-II, respectively) were obtained in a manner similar to the synthesis of ChMAA-n. The melting point of HMAA-II was 58°C and that of BMAA-II, 54°C.

Synthesis of Polymers PMAA-n was obtained by radical polymerization of MAA-n in a solution of dimethyl formamide at 60°C, using dinitrile of azoisobutyric acid (DAA) or UV irradiation at room temperature in an argon atmosphere. The obtained polymers were precipitated with acetone.

Methyl ethers of PMAA-n (PMMAA-n) were obtained by treating PMAA-n with diazomethane in benzene.

ChMAA-n, ChMA, HMAA-II and BMAA-II were polymerized in a benzene solution at 60°C in the presence of DAA. The polymers were precipitated with acetone. Melt polymerization was conducted on the hot stage of a polarizing microscope, in a special cell between cover glasses.

The copolymers of ChMAA-n with A-m and MA-m were obtained by polymerization in benzene in the presence of DAA. The composition of the copolymers was determined by the ratio of optical densities of the 1.740 cm^{-1} (C=O modes in an ester group) and 1.650 cm^{-1} (C=O modes in an amide group) absorption bands. The results of turbidimetric titration of polymers in benzene indicate that the copolymers are homogeneous in composition.

Investigation Techniques Optical studies were conducted in the crossed polarizers of a MIN-8 polarizing microscope with a hot stage. Pictures were taken by means of a "Zenit-3M" camera mounted on the microscope tube via a micro attachment. The X-ray patterns were obtained on a URS-55 X-ray apparatus with a flat cassette (irradiation with CuK_{α}). Small-angle X-ray patterns were obtained on a specially designed camera with a temperature attachment. The sample film distance was adjusted within 90 to 130 mm. Thermographic studies were conducted on a "Derivatograph" instrument (Hungary).

The glass and fusion temperatures were determined from thermomechanical curves derived on a Kargin balance. The IR absorption spectra were obtained on a UR-10 spectrophotometer from samples made in the form of films or pellets with KBr. The turbidimetric titration of copolymer solutions was conducted with methanol. The optical density was measured on a FEK-I photoelectric colorimeter.

Results and Discussion

Monomers Consider first the behaviour of synthesized monomers of the ChMAA-n series at varying temperatures, as well as the effect of crystallization conditions on their structure.

The most active monomers are ChMA and ChMAA-2 which polymerize rapidly at elevated temperatures. For example, melting of ChMAA-2 at 125°C causes the monomer to polymerize immediately, which is indicated by a sharp exothermal peak, on the thermogram of ChMAA-2, following the endothermal melting peak (Figure 2).

In the case of melting of ChMAA-5,6,8,10 monomers, an isotropic liquid is formed at temperatures of 108, 98, 85 and 84°C, respectively. Rapid cooling of these melts results in the formation of liquid crystalline phase whose temperature existence interval depends on the rate of cooling. At slow cooling, either growth of solid crystals (ChMAA-10) or polymerization of monomers (ChMAA-5,6,8) is observed.

ChMAA-II forms two crystalline modifications differing in their melting points and structural parameters (Table III).

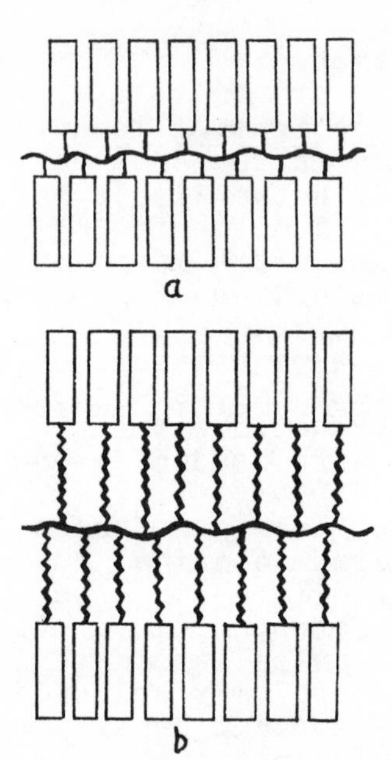

Figure 1. Structure of macromolecules with side mesogenic groups. (a) Mesogenic groups are directly linked to the main chain; (b) mesogenic groups are linked to the branches of comb-like polymers.

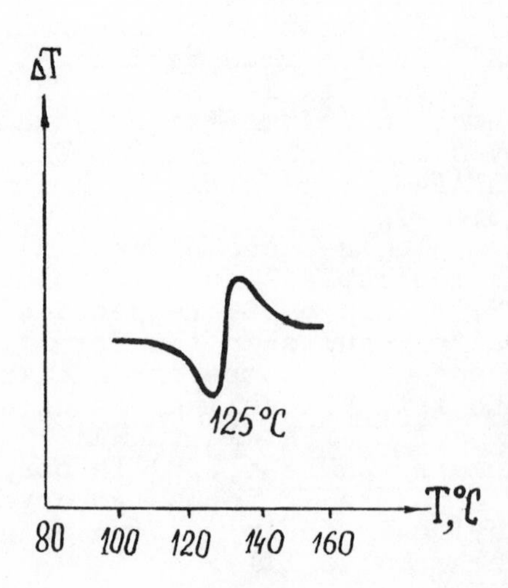

Figure 2. Thermogram of ChMAA-2

Table III. Interplanar spacings and melting points of ChMAA-II samples in various crystalline modifications

Modification	Interplanar spacings, Å					T_m °C
I	4.67	5.10	5.95	7.70	36.0	84
II	3.72	4.90	5.30	7.70	36.0	102

Modification I is formed in the case of crystallization from a solution and, at 84°C, changes to modification II which melts into an isotropic liquid at 102°C. Cooling of the ChMAA-II melt leads to the formation of an liquid crystalline phase which, after further cooling, converts to crystalline modification II. Sequential recording of this process by a camera shows how crystalline spherulites grow in the flowing liquid crystalline phase (Figure 3a) and gradually fill the entire field of vision of the microscope (Figure 3b-d).

Thus, all monomers of the ChMAA-n series form a monotropic liquid crystalline phase of the cholesteric type, whose temperature interval of existence depends on the rate of cooling. The liquid crystalline phase is unstable and is transformed to crystal phase so soon that X-ray examination of the mesophase structure becomes difficult. Nevertheless, polarization-optical studies have made it possible to draw certain conclusions as to the nature of the liquid crystalline phase of monomers. Cooling of isotropic melts of monomers results in a confocal texture which turns to a planar one when a mechanical field is superimposed on the sample, for example, by shifting a cover glass in the cell of the polarizing microscope (Figure 4). The observed planar texture exhibits the property of selective light reflection, which is typical of low-molecular cholesteric liquid crystals.

Polymers and Copolymers Polymerization of all ChMAA-ns and ChMA in a melt yields polymers featuring spontaneous optical anisotropy. The optical pattern observed in crossed polarizers is similar to the confocal texture of low-molecular liquid crystals and represents a combination of birefringent regions 2 to 10 microns is size (Figure 5). At the same time, the presence of a diffuse halo at wide angles of X-ray scattering (Figure 5), in the case of polymers of the PChMAA-n series, does not give reason enough to ascribe crystalline structure to them.

Consider now the manner in which the physical state of a polymer and the optical pattern vary with

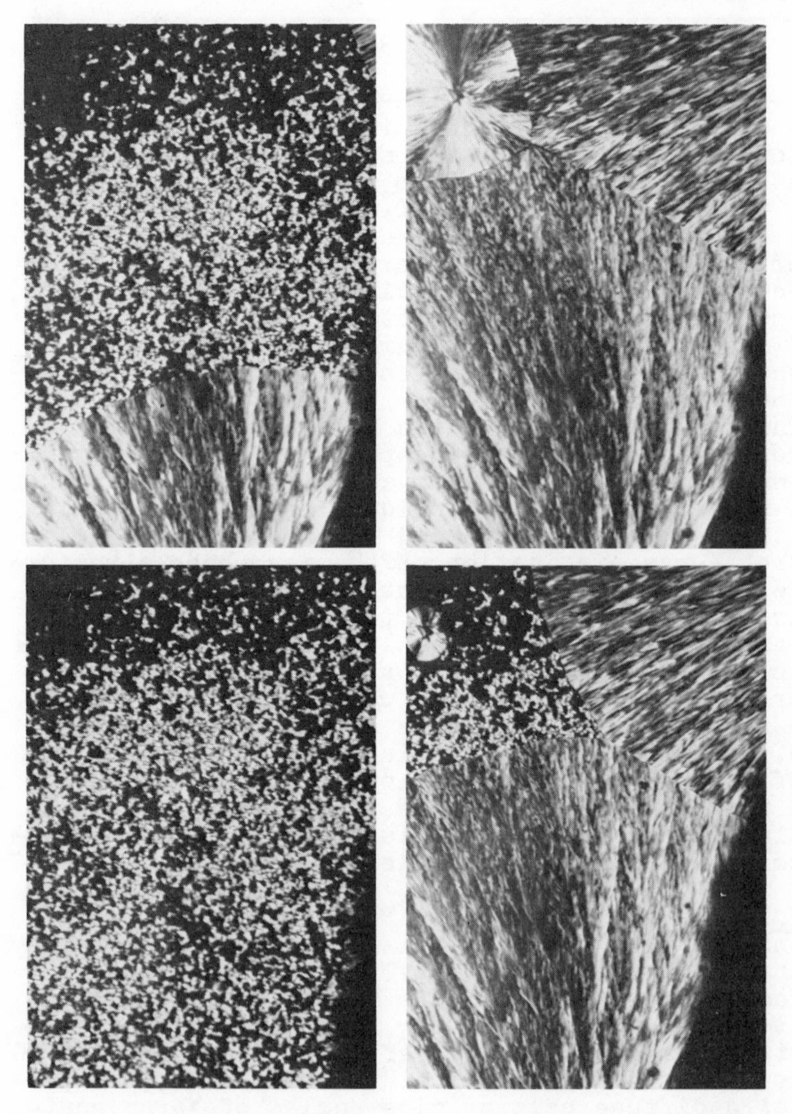

Figure 3. Optical microphotographs showing sequential growth of solid spherulites (b,c,d) from liquid crystalline phase of ChMAA-II (a) (crossed polarizers)

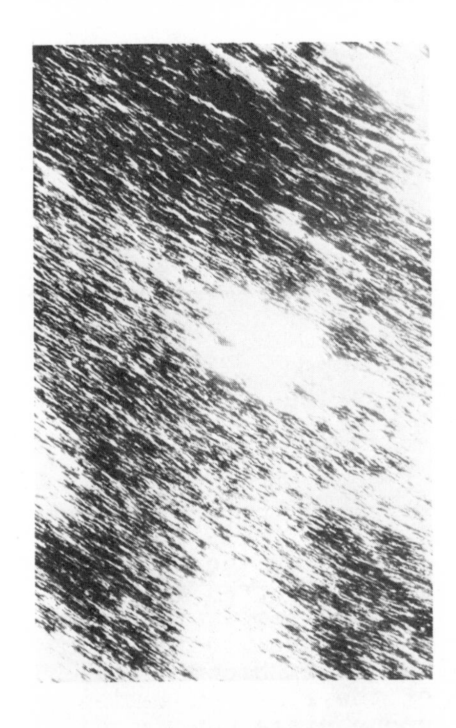

Figure 4. Optical microphotograph of the planar texture ChMAA-II (crossed polarizers)

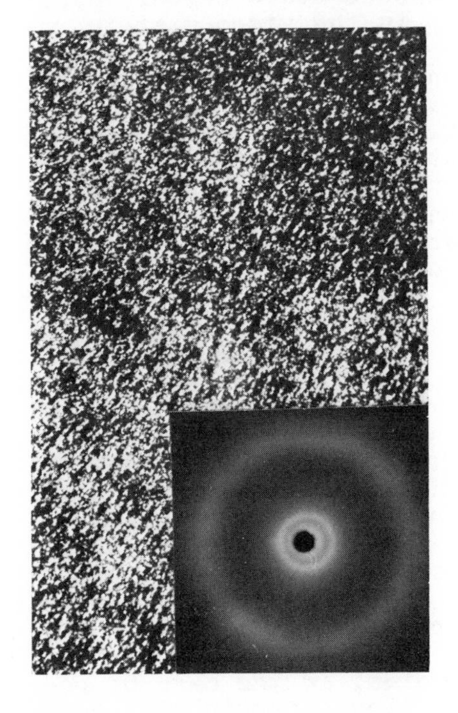

Figure 5. Optical microphotograph of a PChMAA-II film at 20°C. The x-ray pattern of the PChMAA-II film, taken at 20°C, can be seen in the lower right corner.

temperature (Table IV).

Table IV. Glass temperatures (T_g), flow temperatures (T_f) and temperatures^g of transition of from anisotropic to isotropic state ($A_{a\to i}$) for PChMAA-n

Polymer	T_g, °C	T_f, °C	$T_{a\to i}$, °C
PChMA	> 200	-	-
PChMAA-2	185	-	-
PChMAA-5	130	200	220
PChMAA-6	130	190	215
PChMAA-8	130	180	200
PChMAA-10	125	150	185
PChMAA-11	120	135	180

As can be seen from this table, PChMAA-2 has the highest glass temperature (T_g) at which the polymer starts to thermally decompose. The observed optical pattern remain invariable up to this temperature. As the methylene bridge linking cholesterol to the main chain becomes longer, the values of T_g become lower, and, in the case of polymers of the PChMAA-n series with $n \geqslant 5$, an elastic and viscous state may prevail. It should be emphasized that, in the viscous state region, polymers of the PChMAA-n series are viscous fluids whose flow is accompanied by a displacement of the birefringent regions in a way similar to the behavior of low-molecular liquid crystals. As can be inferred from Table IV, the optical anisotropy remains in the elastic and viscous states and disappears at temperature $T_{a\to i}$. Transition from the optically anisotropic to an optically isotropic state at $T_{a\to i}$ is reverse and occurs in a narrow temperature range (2 to 3°).

Table IV shows that the values of transition temperatures $T_{a\to i}$ slightly decrease with increasing n, which is probably due to the effect of internal plastification and was frequently observed in the case of comb-like polymers (21). The evaluation of the thermal effect inherent in this transition for PChMAA-II gave the value of 0.76 ± 0.08 cal/g, which agrees well with the values of heats of melting, corresponding to the liquid crystalline phase-isotropic melt transition for low-molecular liquid crystals (25). However, in contrast to the latter, which, as a rule, crystallize during cooling (see, for example, Figure 3), the liquid crystalline phase of polymers vitrifies while cooled (Figure 5). In other words, in the case of polymers , the structure of the liquid crystalline phase remains in the glass state as well, whereas polymers of the PChMAA-n series exhibit liquid crystalline properties in all three physical states: vitrified, elastic and viscous. The upper limit temperature range of the liquid crystalline state is $T_{a\to i}$. What is it then that makes the liquid crystalline state possible in the polymers under investigation? To answer this question, let us analyze the X-ray study results.

In the wide angles of X-ray scattering there is a diffuse maximum d_I whose magnitude, in the case of $n \geqslant 5$, does not depend on the branch length (Figure 6). In the region of small angles, three maxima are observed whose position changes with the number of the methylene groups linking cholesterol to the main chain (Figure 6).

To interprete these X-ray spacings, let us first examine the structure of model compounds: PMAA-n and PMMAA-n. The macromolecules of these polymers have a chemical structure similar to that of PChMAA-n with the difference that they do not contain cholesterol groups in the side chains.

The X-ray patterns of PMAA-n and PMMAA-n in the region of large scattering angles also display a diffuse maximum corresponding to interplanar spacings: d_I = 4.6 to 4.7 Å, while in the small-angle region a single X-ray spacings is observed, whose position depends on the branch length (Figure 6). The X-ray spacings in the wide angle region is similar to that observed on the X-ray patterns of amorphous poly-n-alkylacrylates and poly-n-alkylmethacrylates, where it is related to the side groups interaction (21).

Evidently, both in the case of PMAA-n, PMMAA-n and PChMAA-n, the X-ray spacings in the wide angle region may be attributed to the distance between the branches arranged in parallel. This assumption is substantiated by the fact that, in the case of uniaxial orientation of polymers, the intensity of this X-ray interference in a meridional direction sharply increases (Figure 7). The value of d_I is 6.3 Å for PChMA and PChMAA-2, while for the other polymers of the PChMAA-n series it is 5.9 Å. These values are slightly greater than the respective interplanar spacings for PMAA-n and PMMAA-n (which, in our opinion, is due to the presence of bulky cholesterol groups) and are close to the distances between the molecules of low-molecular cholesterol esters in the cholesteric mesophase (26). The reason why the values of d_I differ in PChMA and PChMAA-2, on the one hand, and in the other PChMAA-ns, on the other, will be considered below when we shall discuss a polymer structure model.

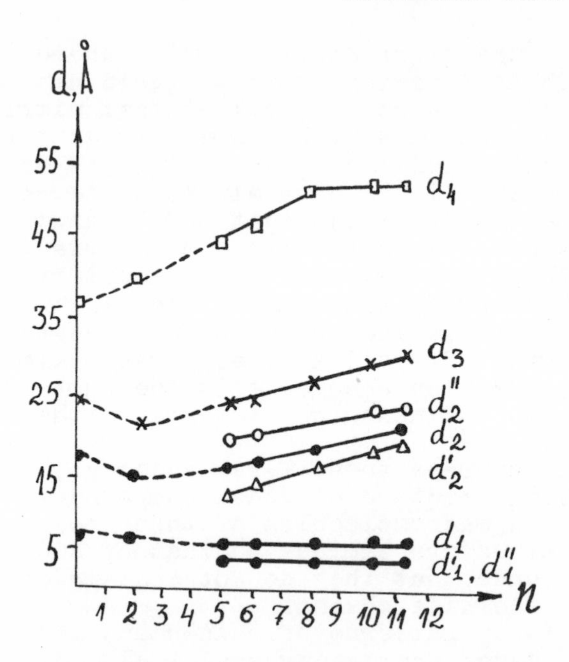

Figure 6. Interplanar spacings vs. the number of carbons in the side chain for PChMAA-n (d_1–d_4), PMAA-n (d_1', d_2') and PMMAA-n (d_1'', d_2''). The experimental points on the ordinate correspond to the interplanar spacings for polycholesterylmethacrylate.

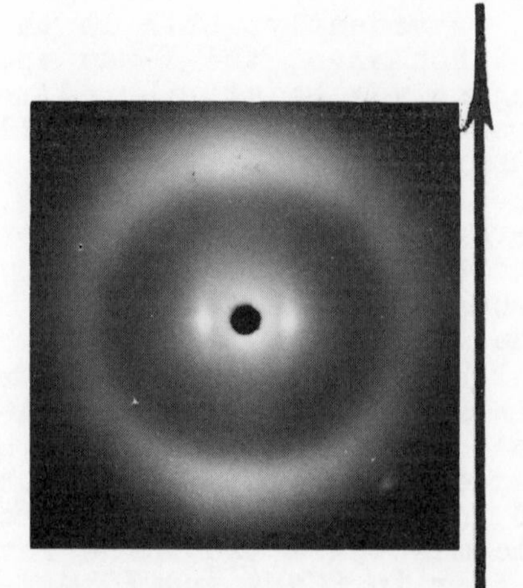

Figure 7. X-ray pattern of an oriented PChMAA-II sample. The primary beam extends at a right angle to the fiber axis.

As to small-angle X-ray spacings, the dependence of spacings d_2 (PMAA-n, PMMAA-n, PChMAA-n) and d_3 (PChMAA-n) on the branch length at $n \geqslant 5$ suggests that they are due to the layered ordering of the side groups. This is corroborated by the X-ray texture patterns of oriented polymer samples. In the case of uniaxial orientation, these X-ray interferences look like sagittal arcs (Figure 7), which is indicative of the branches being arranged substantially at a right angle to the main chain axis. Refraining, at this point, from final and unambiguous interpretation of X-ray spacings d_4, we shall only note that its magnitude is little dependent (and at n = 8,10 and 11, not dependent at all) on the branch length, attributing this interference to the scattering from the main chain and assuming that it is responsible for a certain periodicity associated with its folded structure, if such occurs.

On the basis of the obtained X-ray data, the structure of cholesterol-containing polymers of the PChMAA-n series may be presented as follows.

First of all, it is evident that the structure of PChMA and PChMAA-2 differs from that of PChMAA-n where $n \geqslant 5$. This difference manifests itself both in the values of d_I and the dependence of small-angle X-ray spacings d_2, d_3, and d_4 on n.

In the case of polymers of the PChMAA-n series, where $n \geqslant 5$, d_I = 5.9 Å. This value approaches that for the cholesteric mesophase of cholesterylnonanoate (6.05 Å) and cholesterylmyristate (5.73 Å) (26). Therefore, the packing of branches in PChMAA-n seems to be similar to that of molecules of the above-mentioned low-molecular cholesterol esters in the cholesteric mesophase. Wendorff and Price (26) who had studied the mesophase structure of these compounds suggested that the packing of molecules in the cholesteric mesophase must be antiparallel so that a cholesterol group is surrounded by methylene "tails", while the latter are surrounded by cholesterol groups. This is confirmed by the fact that the lateral size of a cholesterol group exceeds 6 Å, therefore, in the case of parallel arrangement of molecules, d_I would be greater than 6 Å. From the same considerations, it can be assumed that the packing of side chains in polymers of the PChMAA-n series, where $n \geqslant 5$, must be antiparallel so that the cholesterol groups of one macromolecule are surrounded by the methylene chains of adjacent macromolecules (Figure 8a-c). In this case, the branches are arranged at a right angle to the main chain and almost in parallel to one another, however, the long axes of the cholesterol-containing branches may

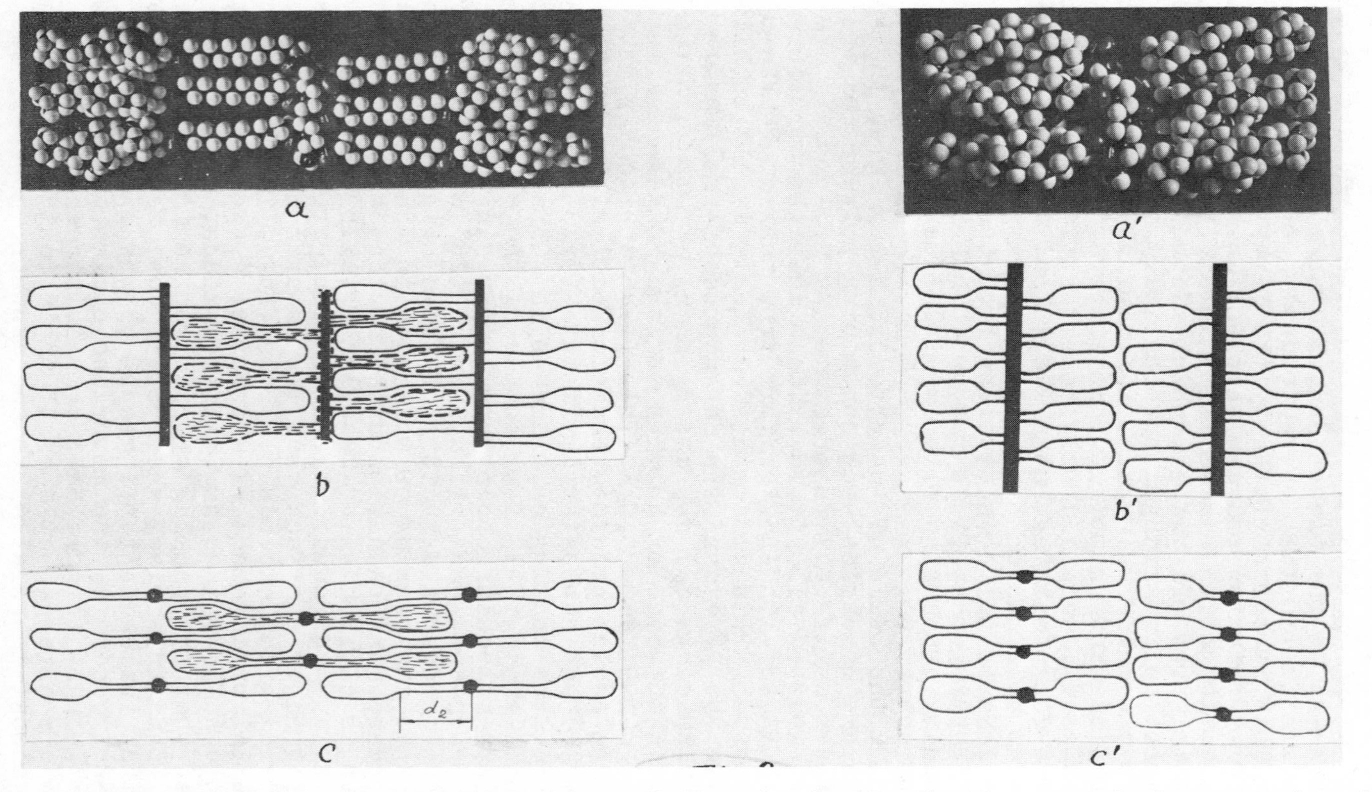

Figure 8. Models of PChMAA-II (a) and PChMA(a') macromolecules and packing of PChMAA-n macromolecules in oriented samples (b,b',c,c') at n $\geqslant$ 5 (b,c) and n < 5 (b',c'); b, b'—solid lines indicate macromolecules lying in the drawing plane, broken lines indicate macromolecules lying in a plane parallel to that of the drawing; c,c'—projection along the main chain of macromolecules.

extend at a certain angle to one another.

As can be seen from Figure 8a-c, the proposed structural model may occur only in such cases in which the length on the sequence of methylene units permits the "rigid" cholesterol skeleton to extend along the methylene chain. Calculations and studies using molecular models suggest that the length of a methylene "bridge" must be at least 10-11 Å. In the case of polymers of the PChMAA-n series, this is true at $n \geqslant 5$. At lower values of n, the cholesterol group cannot extend along the methylene chain, therefore, parallel packing of branches takes place (Figure 8a'-c'). This results in d_I increasing to 6.3 Å, which is close to a respective value for cholesterylacetate (6.48 Å) and for which no antiparallel packing is possible either.

Another reason why it is difficult to discuss the structural model of cholesterol-containing polymers is the fact that so far no adequately substantiated structural model of low-molecular cholesteric liquid crystals exists, apart from the already mentioned packing pattern proposed in Ref. (26). It seems that for a full understanding of the mesophase structure of cholesterol esters, one should, first of all, carefully study their crystalline structure. Such a study, described in Refs. (27, 28), has not yet lead to a complete description of molecular packing in cholesterol esters.

Although the structure of cholesterol-containing polymers is not yet completely understood, there is no doubt that the liquid crystalline properties exhibited by these polymers are due to a particular order in the arrangement of cholesterol groups. This is corroborated by the results of studying the structure and optical properties of model compounds not containing cholesterol. The above-mentioned PMAA-n and PMMAA-n are optically isotropic in all three physical states: vitrified, elastic and viscous.

To elucidate the role of cholesterol in the formation of a liquid crystalline structure, consider now the structure and properties of model PBMAA-II and PHMAA-II polymers whose macromolecules lack cholesterol groups. At room temperature, PBMAA is in elastic state and exhibits no optical anisotropy. On the X-ray pattern of this polymer in the small-angle region, one can see a reflex corresponding to an interplanar spacing of 28 Å. Substitution of the hexadecyl radical for cholesterol results in a crystalline PHMAA-II polymer with a melting point of 40°C. The polymer is crystallized in a hexagonal cell, which is indicated by an intensive X-ray interference at wide angles, corresponding to an interplanar spacing of 4.19 Å.

In the region of small angles, three X-ray interferences are observed, corresponding to interplanar spacings d_2= 14.5 Å, d_3 = 21.6 Å, and d_4 = 38 Å. Above the melting point, the interference in the wide angle region becomes an amophous halo corresponding to an interplanar spacing of 4.6 Å, while in the small-angle region there remains a single interference (d_2 = 36 Å) corresponding in magnitude to the branch length. At temperatures above the melting point, the polymer is optically isotropic.

The foregoing data indicate that in PMAA-n, PMMAA-n, PBMAA and PHMAA-11, the branches are arranged in a layered manner. However, the layered ordering alone is not sufficient for manifestation of optical anisotropy. A liquid crystalline state may occur only in the presence of groups capable of producing mesomorphous structures. In cholesterol-containing polymers, it is precisely the cholesterol groups that perform this function, their mutual packing being responsible for optical anisotropy.

We have already considered the structure and properties of polymers containing cholesterol in such monomer unit. It is also of interest to examine the structure and properties of polymers in which the content of cholesterol groups can be varied, i.e., copolymers. Given in Table V are transition temperatures for some copolymers that we have synthesized.

Table V. T_g, T_f and $T_{a\to i}$ of copolymers of ChMAA-n with alkylacrylates (A-m) and alkylmethacrylates (MA-m)

Copolymer (mol.%, ChMAA-n)	T_g, °C	T_f, °C	$T_{a\to i}$, °C
ChMAA-II with A-4			
42	65	115	160
37	60	100	140
17	< 20	60	100
ChMAA-II with MA-4			
90	115	140	180
67	105	140	170
40	85	135	160
ChMAA-II with MA-10			
75	90	140	180
58	70	120	170
25	< 20	90	no anisotropy
ChMAA-II with A-16			
45	45	70	100
ChMAA-II with MA-22			
75	70	110	no anisotropy

50	40	80	no anisotropy
ChMAA-6 with A-4			
45	100	170	180
30	70	105	115

As can be seen from this table, the introduction of a second component, namely, a "solvent" of cholesterol-containing units, whereby the glass temperature is lowered, resulted, in some cases, in a longer interval of existence of the elastic and viscous liquid-crystalline states, while in other cases, in the formation of optically isotropic polymers.

Copolymers of ChMAA-II with butylacrylate (A-4) and butylmethacrylate (MA-4) form a liquid crystalline phase in a wide range of component ratios (Table V and Figure 9a,b). For example, a copolymer containing more than 80% of A-4 can still exist in the liquid crystalline state. At the same time, a copolymer with MA-22 whose content is only 25%, is amorphous and optically isotropic. As had been mentioned above, the optical anisotropy in cholesterol-containing polymers is due to a particular ordering in the arrangement of the cholesterol groups. If, in copolymers, the second component does not hinder packing of the cholesterol groups, a mesophase may be formed. In copolymers with MA-22, the long branches screen the cholesterol groups, and no optical anisotropy is manifest in any of the three physical states of polymers.

It should be noted that a copolymer with an equimolar ratio of ChMAA-II and MA-22 units is amorphous, too (Table V), although the PMA-22 homopolymer easily crystallizes, forming a hexagonal lattice typical of comb-like polymers (29). As was shown earlier (21), the introduction of such "disturbers" as isopropylacrylate in an amount of 80% into crystalline comb-like polymers does not destroy the crystal lattice. In our case, the introduction of only 50% of ChMAA-II into PMA-22 prevents the latter from crystallizing. This copolymer provides a striking example of the impossibility of a structure characteristic of each of the homopolymers in the case of copolymerization of monomers with bulky groups.

Thus, copolymers of cholesterol-containing monomers with n-alkylacrylates and n-alkylmethacrylates are capable of yielding a liquid crystalline phase in cases where the length of the alkyl chain in alkylacrylate or alkylmethacrylate does not exceed that of the methylene bridge linking cholesterol with the main chain.

An X-ray study of copolymers has shown that the introduction of a second component into cholesterol-containing polymers results in a certain change in the latter's structure. In the small-angle region, only one diffraction maximum is retained for all copolymers ChMAA-II, corresponding to interplanar spacing d = 19.8 Å which coincides with d_2 for the PChMAA-II homopolymer. In the case of uniaxial orientation, this X-ray interference takes the form of sagittal arcs. Since copolymers of ChMAA-II are optically anisotropic, one naturally assumes that it is precisely the high degree of ordering in the arrangement of the side methylene chains, manifesting itself in the occurrence of interference d_2, that permits such packing of cholesterol groups, which ensures the optical anisotropy in the above polymers. This is supported by the results of X-ray studies of copolymers at various temperatures. Figure 10 shows that the intensity of the small-angle X-ray spacings remains high in the viscous state as well when there is an optical anisotropy (Figures 9a,b), and sharply decreases during transition to an optically isotropic state (Figure 10, curve 3).

The results of studying the structure and properties of copolymers demonstrate that proper selection of components for copolymerization permits obtaining a wide variety of polymers forming a mesophase in different temperature ranges.

As far as the type of the liquid crystalline phase formed in cholesterol-containing polymers is concerned, the morphological similarity of the texture appearing in polymer films with the confocal texture of cholesteric liquid crystalline compounds seems to suggest that we deal with a cholesteric type of liquid crystals. However, the planar texture typical of cholesteric liquid crystals, which is essentially a single crystal of the cholesteric type, does not occur in films of these polymers, and when a cover glass is shifted, the texture formed in the polarizing microscope cell is deformed (Figure 9b). This is probably due to the defects introduced into the resulting liquid crystalline structure by the polymer main chain. On the other hand, the layered ordering in the arrangement of branches is indicative of a smectic structure, hence, the seeming similarity of the texture of low-molecular liquid crystals of the cholesteric type to polymers of the PChMAA-n series and their copolymers has, in fact, nothing to do with the manner in which structural elements of comb-like macromolecules are packed in the cholesteric phase. Besides, the presence of the main polymer chain hinders the formation of a spiral structure typical of low-molecular cholesteric liquid crystals. All this

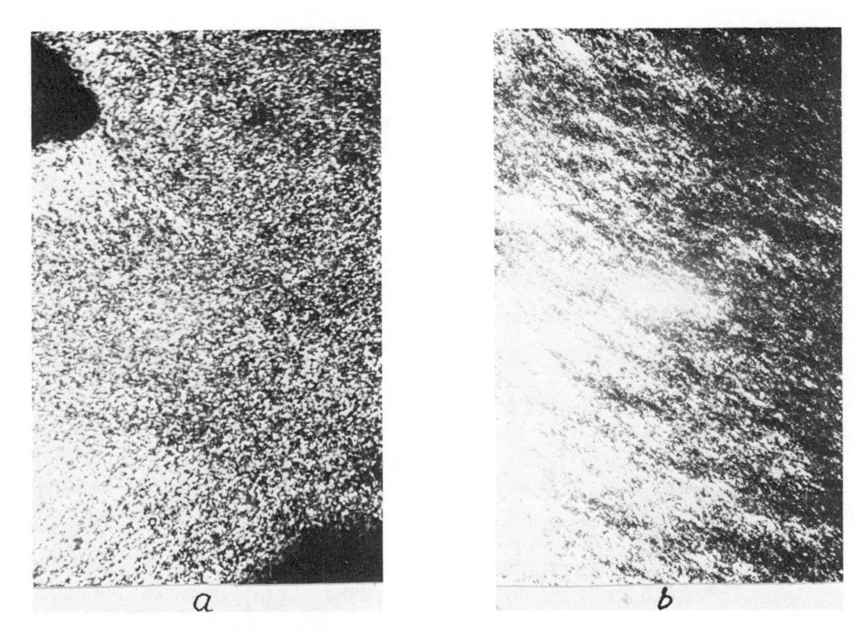

Figure 9. Optical microphotographs of a copolymer of ChMAA-II with A-4 (37:63), at 130°C without superposition of a mechanical field (a) and after shifting a cover glass in the microscope cell (b) (crossed polarizers)

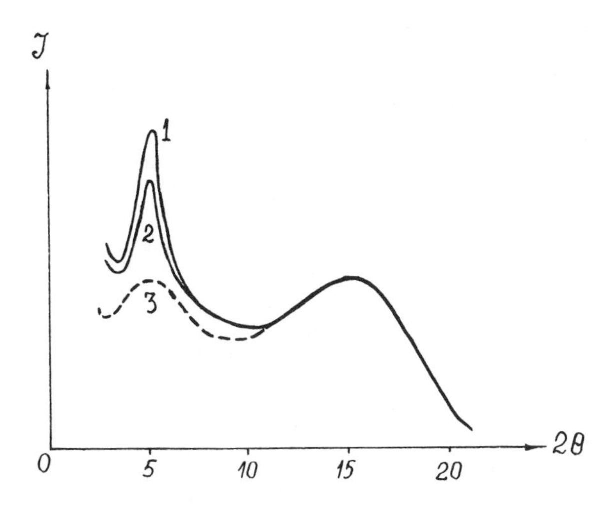

Figure 10. X-ray scattering intensity vs. scattering angle for a copolymer of ChMAA-II with A-4 (37:63) at 25 (1), 130 (2) and 150°C (3)

renders it difficult to class the liquid crystalline phase of polymers with a definite type of liquid crystals. Evidently, the classification of liquid crystals proposed for low -molecular compounds cannot be applied to polymers whose structural arrangement is different from that of low-molecular liquid crystalline systems.

We wish to acknowledge the valuable collaboration of I.V.Sochava whose group at the Physical Institute of the Leningrad State University has determined the melting heat of the liquid crystalline phase of PChMAA-II.

Literature cited

1. De Viseer, A.C., De Groot, K., Feyen, J., and Bantjes, A., J.Polymer Science, (1971), A-1, 9, 1893.
2. Bouligand, J., Cladis, P., Liebert, L., and Strzelecki, L., Mol.Cryst.Liq.Cryst., (1974) 25, 233.
3. Amerik, Yu.B., and Krentsel', B.A., in coll."Uspekhi khimii i fiziki polimerov" (Advances in Chemistry and physics of Polymers), p. 97, "Khimiya" Publishers, Moscow, 1973.
4. Frenkel, S.Ja., J.Polymer Sci. (1974), C-44, 49.
5. Blumstein, A., Blumstein, R., Murphy, G., Wilson, C., Billard, J., "Liquid Crystals and Ordered Fluids", p. 277, vol.2, Plenum Press, 1974.
6. Freidzon, Ya., Shibaev, V.P., and Platé, N.A., Abstracts of papers at the 3rd All-Union Conference on Liquid Crystals, p. 214, Ivanovo, 1974.
7. Shibaev, V.P., Talrose, R.V., Karahanova, F.J., Haritonov, A.V., and Platé, N.A., Dokl.AN SSSR (1975), 225, 632.
8. Roviello, A., and Sirigu, A., Polymer.Lett.Ed. (1975), 13, 455.
9. Perplies, E., Ringsdorf, H., and Wendorff, J., Makromolek.Chemie (1974), 175, 553.
10. Lorkowski, H., and Reuther, F., Plaste und Kautschuk, (1976), 23, 81.
11. Shibaev, V.P., Freidzon, Ya.S., and Platé, N.A., Dokl.AN SSSR (1976), 227, 1412.
12. Papkov, S.P., Vysokomolek.soed. (1977), A-19, 3.
13. Shibaev, V.P., and Platé, N.A., Vysokomolek.soed. (1977), A-19, 923.
14. Shibaev, V.P., Freidzon, Ya.S., Platé, N.A., Vysokomolek.soed., (1978), A-20, in press.
15. Krentsel', B.A., and Amerik, Yu.B., Vysokomolek. soed. (1971) A-13, 1358.
16. Blumstein, A., Blumstein, R., Clough, S., and Hsu, H., Macromolecules (1975), 8, 73.

17. Clough, S., Blumstein, A., and Hsu, E., Macromolecules (1976), 9, 123.
18. Blumstein, A., Clough, S., Patel, L., Blumstein, R., and Hsu, H., Macromolecules (1976), 9, 243.
19. Shibaev, V.P., Freidzon, Ya.S., and Platé, N.A., Abstracts of papers at the 11th Mendeleev Congress on General and Applied Chemistry, p. 164, vol. 2, "Nauka" Publishers, (Moscow), 1975.
20. Shibaev, V.P., Freidzon, Ya.S., and Platé, N.A., USSR Inventor's Certificate No. 525, 709, Byull. izobreteniy (1976), No. 31.
21. Platè, N.A., and Shibaev, V.P., J.Polymer Sci.,-Macromolec.Rev., (1974), 8, 117.
22. Kamogawa, H., J.Polymer Sci. (1972), B-10, 7.
23. Minezaki, S., Nakaya, T., and Imoto, M., Makromolek.Chemie, (1974), 175, 3017.
24. Saeki, H., Iimura, K., and Takeda, M., Polymer.J. (1972), 3, 414.
25. Kunihisa, K., and Hagiwara, S., Bull.Chem.Soc. Japan, (1976), 49, 2658.
26. Wendorff, J., and Price, F., Mol.Cryst.Liq.Cryst. (1973), 24, 129.
27. Wendorff, J., and Price, F., Mol.Cryst.Liq.Cryst. (1973), 22, 85.
28. Barnard, J., and Lydon, J., Mol.Cryst.Liq.Cryst., (1974), 26, 285.
29. Shibaev, V.P., and Freidzon, Ya.S., Vysokomolek. soed., (1975), B-17, 151.

RECEIVED December 8, 1977.

5

Liquid Crystalline Order in Polymers and Copolymers with Cholesteric Side Groups

A. BLUMSTEIN, Y. OSADA,[1] S. B. CLOUGH, E. C. HSU, and R. B. BLUMSTEIN

Department of Chemistry, Polymer Program, University of Lowell, Lowell, MA 01854

We have recently shown (1, 2) that atactic polymers with side groups characterized by chemical constitution related to mesomorphic behavior tend to organize spontaneously and independently of the conditions of their preparation if given enough segmental mobility. The organization displayed can be lamellar (smectic) or directional (nematic). In the former case the macromolecular backbone is confined to a plane, while the side groups are either perpendicular or tilted with respect to the planes containing the backbone. The side groups are arranged either in single or in double arrays in which the individual side groups are mutually parallel (3). In the latter (nematic) type of organization, the backbones are not restricted to layers but the side groups keep a mutually parallel orientation throughout the domain which can reach submicroscopic dimensions.

The development of liquid crystalline order in the polymer may or may not be accompanied by crystallinity. Polymers with benzoic acid side groups were studied and found to display ordering analogous to the smectic C structures of alkoxy benzoic acids (1, 2). Polymers with Schiff base side groups were found to display ordering of the smectic A (1) and nematic type (4). Literature X-ray data (5,6) can be interpreted as being generated by lamellar (smectic) structures and seem to point to the fact that in all cases the intrinsic tendency of side groups to organize is reinforced by the connectivity imposed by the polymeric backbone.

Mesomorphic order in polymers with side groups containing the cholesterol moiety has scarcely been studied in spite of the importance of cholesterol derivatives in the field of liquid crystals and biological materials. The main focus of attention has been

[1] Present address: College of Liberal Arts and Science, Ibaraki University, Mito 310, Japan.

0-8412-0419-5/78/47-074-056$05.00/0

centered around the study of phase transitions of acrylic (7, 8, 9) and methacrylic (10, 11, 12) esters of cholesterol. In spite of considerable controversy as to the location of the cholesteric phase transition in these monomers, there was a consensus that polymers of cholesterylacrylate and cholesterylmethacrylate are amorphous. The amorphous structure is reported to occur independently of the phase in which the polymerization has been carried out, though Hardy (7) has shown that polymer-monomer mixtures (polymerization at $0^{\circ}C$) have a smectic organization. From our point of view, these statements were surprising as we expected the cholesteric moieties to interact strongly and organize the polymer in a fashion similar to other polymers with mesogenic side groups. They warranted an inquiry into the structure of polymers with cholesteric side groups.

A recent publication (13) which appeared while this work was in progress described the spontaneous development of layered order in polymers with the cholesteric moiety attached to the backbone via a long, flexible molecular spacer constituted by an ω-aminocarboxylic acid chain. Layered structures are developed with the long side groups roughly perpendicular to the main chain. The authors attributed the development of these structures to the flexibility of the bridge connecting the cholesterol moiety.

One of the factors leading to confusion in interpreting data on the development of order in polymers with mesogenic side groups is the frequent phase separation taking place during the polymerization process. The intermolecular disorder characterizing the precipitated polymer is often "locked in" if the temperature of precipitation is below the glass transition temperature of the polymer Tg. Casting of films from good, slowly evaporating solvents or skillful annealing invariably leads to the development of mesomorphic order (1, 3). Similar results have been achieved with plastification of poly(γ-L-benzylglutamate)(14).

In this paper we describe the study of different polymers containing the cholesterol moiety in the side group as well as copolymers of cholesterylmethacrylate (PChMA) with n-alkylmethacrylates of various chain lengths. The development of order in PChMA's of different tacticities is also described, providing some insight into the factors governing the development of mesomorphic order in polymers with cholesteric side groups.

Experimental

All solvents and reagents were distilled prior to use.

Monomers. Cholesteryl p-acryloyloxybenzoate (ChAB) was prepared in three steps by the following synthetic route:

$$HO-C_6H_4-COOH + CH_2=CH-\overset{O}{\overset{\|}{C}}-Cl \xrightarrow[2)\ H^+]{1)\ OH^-} CH_2=CH-\overset{O}{\overset{\|}{C}}O-C_6H_4-COOH$$

$$\xrightarrow{SOCl_2} CH_2=CH-\overset{O}{\overset{\|}{C}}O-C_6H_4-\overset{O}{\overset{\|}{C}}Cl \xrightarrow[C_6H_5-N(CH_3)_2]{cholesterol}$$

$$CH_2=CH-\overset{O}{\overset{\|}{C}}O-C_6H_4-\overset{O}{\overset{\|}{C}}O[Ch]$$

The product was recrystallized from chloroform/acetone and chloroform/ethanol to constant transition temperature. Elemental analysis: Calcd. for $C_{37}H_{52}O_4$: C, 79.29%; H, 9.29%. Found: C, 79.51%; H, 9.31%. The NMR spectra were consistent with the expected structure.

Cholesterylmethacrylate (ChMA) was obtained by a procedure similar to the one described in the literature (15).

Thirty five grams (0.34M) of methacryloylchloride was added slowly to a mixture of 77 g (0.2M) of cholesterol and 150 ml of triethylamine dissolved in 500 ml of toluene. After 3 h another 150 ml of triethylamine was added to the mixture and the reaction was continued for 24 h at 60-70°C. The cholesterylmethacrylate was precipitated three times into 1.6N HCl solution in methanol from toluene. The yield was 60-70% The pure compound displayed a melting point at 114.8°C and a monotropic cholesteric phase at 111.8°C: K 114.8 I (111.8) in agreement with literature data. (K, I and C designate the crystalline, isotropic and cholesteric phase, respectively. The elemental analysis, NMR and IR spectra were consistent with the structure of cholesterylmethacrylate.

Methylmethacrylate (MMA), propylmethacrylate (PMA)

n-butylmethacrylate (BMA), n-octylmethacrylate (OMA), and n-dodecylmethacrylate (DMA), were purified by distillation under vacuum. The n-hexadecylmethacrylate (HMA) was recrystallized from a benzene-ether mixture.

Polymerization and Copolymerization. The ionic polymerization of ChMA was carried out in toluene (dried over Na wire and distilled in presence of $LiAlH_4$), at -30°C using BuLi as initiator. The polymer was precipitated three times from benzene into methanol. The free radical polymerization of ChA and ChMA was carried out thermally or in solution in freshly distilled benzene at 60°C using AIBN as initiator. The total concentration of monomer was 1.0 mol/l. The concentration of initiator was 10^{-2} mol/l. Glass ampules were filled to half capacity with the solution, evacuated, purged with N_2 and sealed. They were placed in a thermostat at 60°C. The polymerization was interrupted at conversions varying from 5 to 15% (by weight). The polymer was precipitated twice from benzene into methanol and dried at 40°C in vacuo. The copolymer compositions were determined from elemental analysis of the copolymer.

Characterization of Monomers and Polymers. Phase transitions of monomers were determined by means of Differential Scanning Caloimetry (DSC) using the Perkin Elmer DSC 1B and thermal polarizing microscopy.

X-ray studies on the samples were performed with a Rigaku wide angle diffractometer (SG 7B) as well as with a low angle Warhus camera. Cu-K_α radiation was used.

Samples of copolymers were examined as pellets made from cast films. (Films were cast from a 1-2% solution in benzene followed by drying at 50°C in vacuo). Several film samples were pressed together in a die (12,000 p.s.i., 3 min. under vacuum of approx. 0.1mm) to form a pellet 1.5mm thick and 10mm in diameter. Samples of homopolymers were examined either as single unpressed films or as pellets. Spacings characteristic of smectic organization (d) were calculated by the Bragg equation. The average distance between side groups (D) was calculated from the equation $D=1.117\lambda/2\sin\theta$) (16) from the halo near $2\theta=15°$.

The sharpness s of the low angle x-ray diffraction peaks is defined as the ratio of the peak height h to the width w at 1/2 h. The position of the low angle peak for the cholesterylmethacrylate was found at $2\theta=2.5°$ a value somewhat higher than the value $2\theta=2.4$Å found for the same polymer in an earlier study (17).

The glass transition temperatures were estimated from deformation tests performed on the very same pellets which were used for x-ray studies. The tests were made by means of a Perkin-Elmer Thermomechanical analyzer TMS-1, calibrated with polystyrene and polymethylmethacrylate. C-13 NMR Spectra were run in deuterated chloroform at 10wt% concentration and at 45°C. A JEOL PFT-100 spectrometer was used (resolution: 1.22 H_z; pulse width 15 μsec; repetition: 1.0sec). Tacticity assignments were made from the analysis of the carbonyl portion of the spectrum.

Transmission electron microscopy was performed on samples of thin films of copolymers and blends of homopolymers by means of JEOL JEM100 electron microscope. The films were stained with O_sO_4 prior to examination.

Results and Discussion

Monomers. Table I gives the transition temperatures as found by means of DSC and polarizing microscopy for the three monomers. The cholesteric transitions of both cholesteryl acrylate and cholesteryl methacrylate were found to be monotropic in agreement with Toth and Tobolsky (8) and deVisser (9). While the exact value of the transition for ChMA is in agreement with deVisser's (9), the value for ChA is at variance with the value from (8) and adds still another transition temperature to the literature of reported values for this monomer. The cholesteryl ester of the acryloyloxybenzoate (ChAB) gives an enatiotropic transition at 128°C (K⟶C). All three monomers display iridescent colors in their cholesteric state. In the solid state ChA and ChAB exhibited an intense x-ray diffraction peak near $2\theta=2^{\circ}$ as well as a number of wide angle lines. In the cholesteric state the low angle peak disappeared. The x-ray diffraction pattern of ChMA in the solid state contained a low angle peak at $2\theta=5.9^{\circ}$ and a number of wide angle lines. This monomer polymerized rapidly after melting and we did not succeed in recording the x-ray diffraction pattern in the cholesteric state. Thus, the liquid crystalline state of ChA and ChAB has a twisted nematic organization. It is very probable that liquid crystalline ChMA is also a twisted nematic.

Table I

Monomers

Name	Structure	Transition Temperature °C
ChA	$CH_2 = CHC(=O)-O[Ch)$	K 125.8 I (124.8 C 91K)
ChMA	$CH_2=C(CH_3)C(=O)-O[Ch]$	K 114.8 I (11.8C)
ChAB	$CH_2=CHC(=O)O-C_6H_4-C(=O)O[Ch]$	K 128 C⟶poly'n

Polymers. PChMA. For the polymer prepared by free radical initiation the very intense x-ray line which corresponds to 35.3Å ($2\theta=2.5°$) shows a smectic type structure form (see Fig. 1b). This structure is considerably weakened by abrupt precipitation of the polymer. The distance from the center of the main chain to the end of the side group is estimated at 18-19Å from molecular models. The predominantly isotactic form of PChMA prepared by ionic polymerization also displays an intense x-ray line which, after annealing, is located at ($2\theta=2.5°$) indicating a lamellar bi-layered arrangement of macromolecules similar to that of the predominantly syndiotactic specimen. For both forms, a halo with a peak at $15.5°=2\theta$ shows that no long range order exists between side groups. The average distance D between side groups is 6.35Å for both types of PChMA.

Table II

	rr	mm	dÅ	DÅ
PChMA (AIBN) (after annealing)	.6	.1	35.3	6.3
PChMA (BuLi) (after annealing)	.1	.6	35.3	6.3

PChAB. An intense and sharp low angle reflexion at 45.2Å and a halo at 6.08Å indicate a smectic type of organization (see Fig. 1a). Here again the distance from the center of the main chain to the end of the

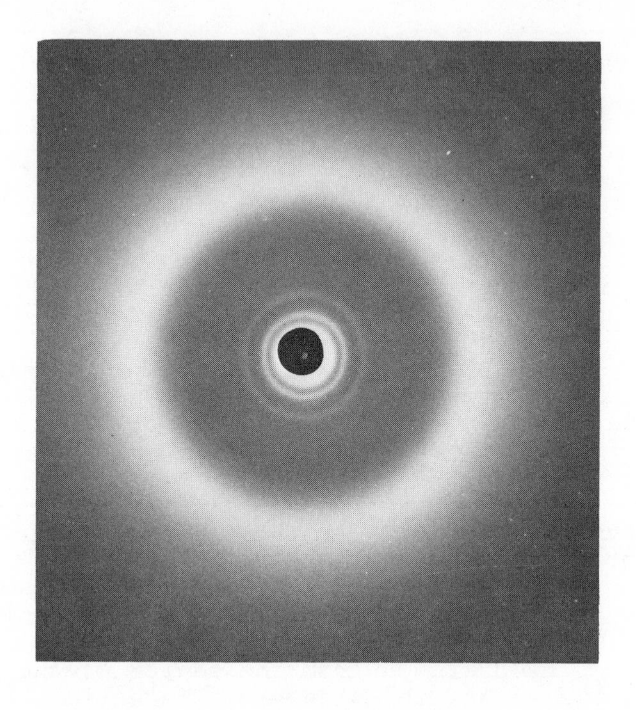

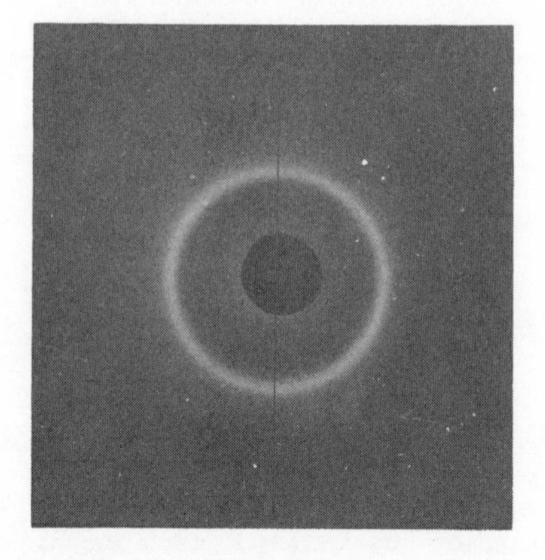

Figure 1. (a) (above) *X-ray low and wide angle diffraction patterns obtained from poly(cholesterylester of the acryloyloxybenzoic acid)(PChAB). (b)*(right) *X-ray low angle diffraction pattern obtained from annealed sample of poly(cholesterylmethacrylate) PChMA.*

side group (23Å estimated from molecular models) is half of the d spacing.

PChA. A broader x-ray peak than for the two preceding polymers is observed near $2.6^{\circ}=2\theta$ (33.4Å). Abrupt precipitation of this polymer leads to an amorphous polymer.

We are inclined to think that in all cases presented above, the main chain of the polymer is confined to planes, with side groups essentially perpendicular to the main chain on both sides of the plane. This model is schematically given in figure 2b.

Copolymers. Figure 3 shows a set of typical low angle diffraction patterns obtained from PChMA and copolymers of various compositions of ChMA and BMA. One can see that the homopolymer as well as some copolymers of high ChMA content are characterized by a sharp low angle x-ray diffraction peak at $2\theta=2.6^{\circ}$ corresponding to 33.9Å, with higher orders (18). The position of this peak does not vary with the composition and nature of the second comonomer, although its intensity is strongly affected. The low angle peak is accompanied by higher orders and a halo around $2\theta=15.5$Å which corresponds to the average distance between cholesteryl side groups of D=6.35Å

The value of 33.9Å for the low angle x-ray diffraction is somewhat lower than 35.3Å found for PChMA in a former study (17). It may well be that, during the sample preparation, the 12,000psi applied on the pellet results in a slight cholesteric side group interpenetration (see Fig. 2a). We have attributed the 33.9Å peak to the regular interlamellar spacings of the ordered polymer. The model proposed for the polymer of ChMA is still applicable to the copolymers with high ChMA content, with the cholesteric side groups arranged in a double array and essentially perpendicular to the lamellar plane containing the convoluted backbone. One must assume that side groups interpenetrate slightly as the length of the side group is 18-19A (estimated from Fisher models from center of the chain to the end of the cholesteric moiety).

Figure 4 gives the sharpness s of the 2.6° peak as a function of the composition for different copolymers of ChMA with n-alkylmethacrylates. These results are compared to the data obtained for mechanical blends of the corresponding homopolymers. The samples of homopolymer blends were prepared under the same conditions as the copolymers.

It is immediately apparent from Figure 4, that depending on the structure of the non-mesogenic co-

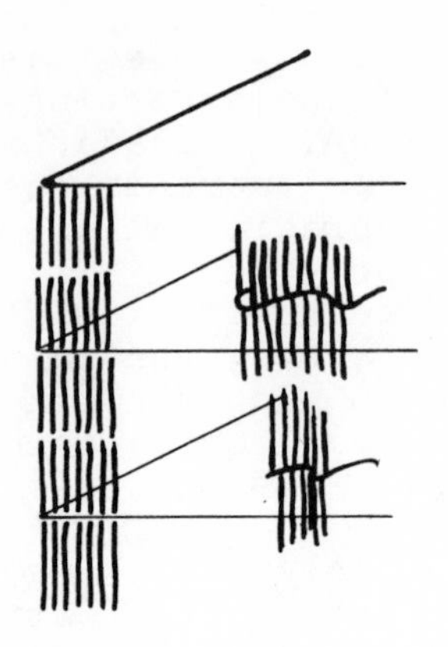

Figure 2. (a) (above) *Cholesteric side groups (Fisher space filling molecular models). (b)* (right) *Proposed smectic organization of PChAB, PChMA,and PChA.*

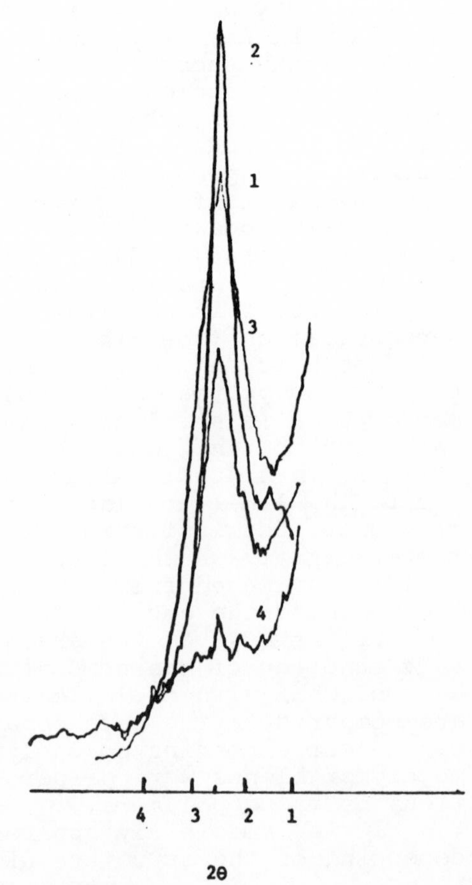

Journal of Polymer Science

Figure 3. Low angle x-ray diffraction peak for PChMA and copolymers of CHMA with BMA of various compositions are expressed in mole %. (1) ChMA (100%), (2) ChMA (91.4%)—BMA (8.6%), (3) ChMA (84.8%)—BMA (15.2%), (4) ChMA (78.1%)—BMA (21.9%) (17).

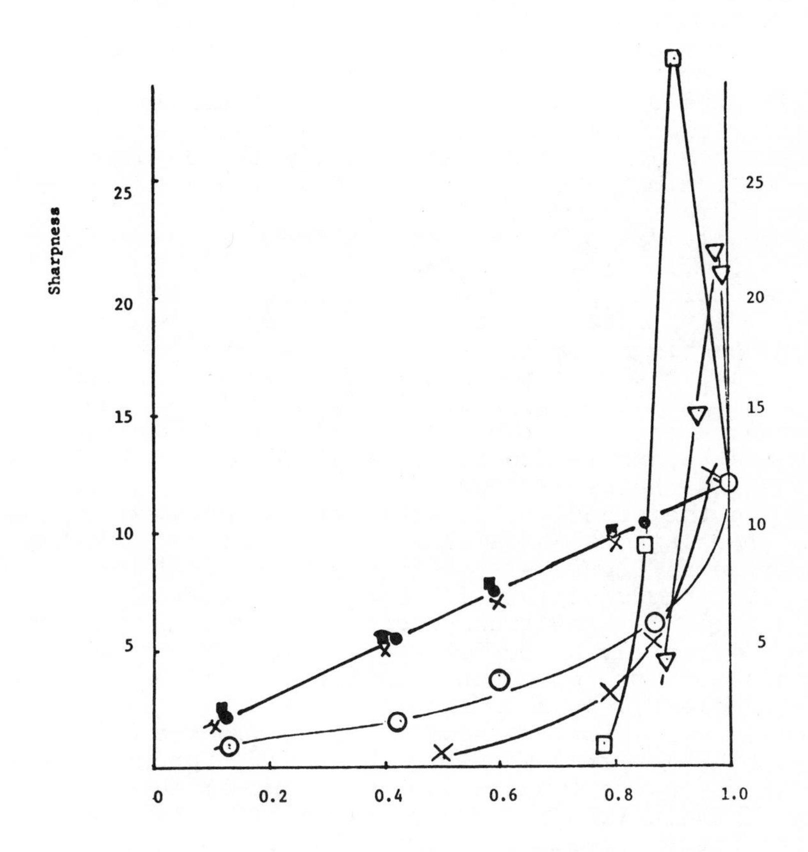

Journal of Polymer Science

Figure 4. (mole %). S as a function of composition in copolymers of ChMA with various long chain methacrylic esters and also, blends of various homopolymers (before annealing). Copolymer: (○) MAA, (×) PMA, (□) BMA, (△) OMA. Blend: (●) MMA, (×) PMA, (■) BMA (17).

monomer, the smectic organization of the copolymer is at first reinforced and then very rapidly declines for high compositions of the non-mesogenic comonomer. This effect is the strongest with BMA but persists for the octylmethacrylate (OMA). In the case of shorter chain esters MMA and PMA the organization is not significantly reinforced and a less precipitous but steady decline of the peak intensity sets in. Quite surprisingly, the copolymer of MMA can still display a significant low angle peak down to less than 40 mole percent of cholesteryl methacrylate in the copolymer. The variation of s with the composition for blends of homopolymers stands in contrast with the results for the copolymers. As one can see from Figure 4, this variation is linear and for a given composition independent of the polymer admixed.

The sharpness s of the low angle x-ray diffraction peak is a complex experimental parameter related not only to the perfection of the lamellar (smectic) domains in the copolymer, but also to their size and concentration. In the case of blends of homopolymers the scattering is due to segregated domains of ChMA units and therefore s would depend on their concentration in the blend. This seems to be confirmed by the linear relationship between sharpness s and blend composition in Figure 4.

The study of thin films of copolymers and blends of homopolymers by transmission electron microscopy gives direct evidence for segregation into separate domains of both components in the case of blends. This can be clearly seen in Figure 5. The O_sO_4 stained cholesteryl moieties show as dark regions and indicate that PChMA separates into clusters of approximately 1μ. No separation was seen in films of the copolymer.

The non-linear and strong variation of s with the composition for all copolymers indicates that the perfection of the lamellar domains is strongly affected by the presence in the copolymer chain of non-mesogenic alkylmethacrylates. The interference takes place already at very low concentrations of the non-mesogenic comonomer. There are good reasons to believe that the drastic increase of s for copolymers of ChMA-BMA and ChMA-OMA is due to an internal plastification effect. Plastification would increase the segmental mobility and give to chains more opportunity to "pack" into a lamellar array. The glass transition temperatures for various copolymers of ChMA and p-alkylmethacrylates are plotted in Figure 6 as a function of composition. One can see that the copolymer compositions corresponding to the strongest increase of s in Fig. 4

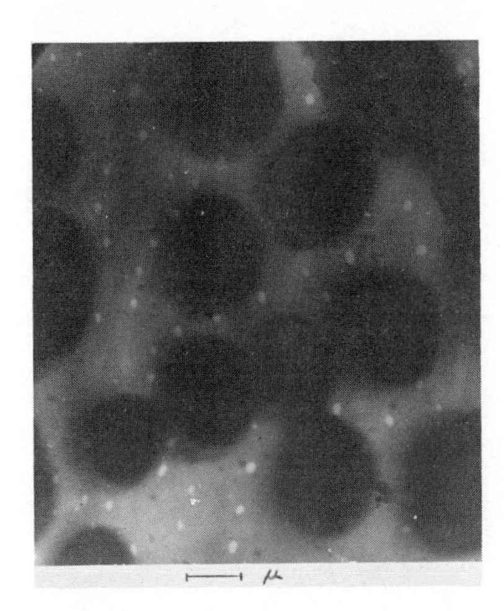

Figure 5. Electron micrograph of a cast film of a blend of PChMA and PBMA. The black regions indicate phase separated domains of PChMA.

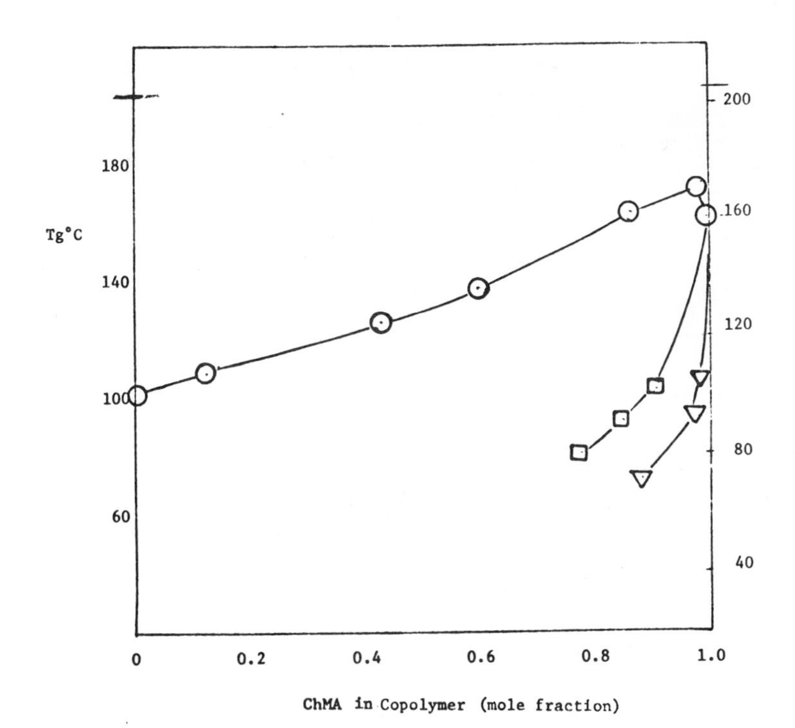

Journal of Polymer Science

Figure 6. Glass transition temperature as a function of composition of the copolymer. (○) MMA, (□) BMA, (△) OMA (17).

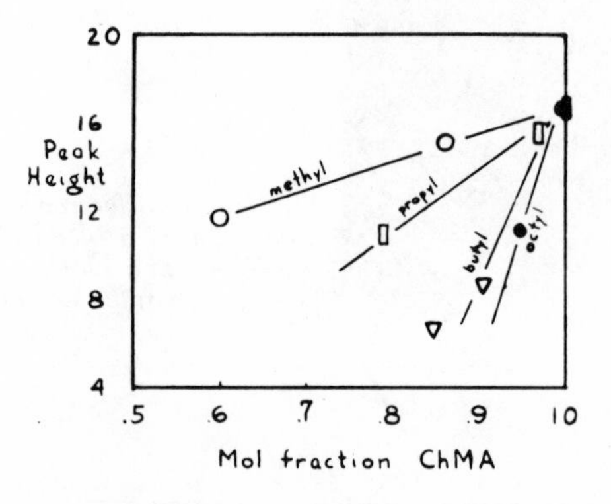

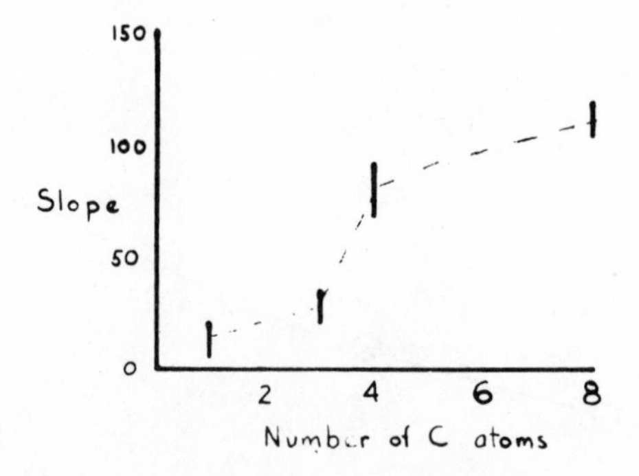

Figure 7. (a) (top) *Height of the low angle x-ray diffraction peak in various copolymers of ChMA as a function of composition after annealing. (b)* (bottom) *Slope of lines from Figure 7a vs. number of carbon atoms in the aliphatic chain of the non-mesogenic methacrylic comonomer.*

give large decrements of Tg. The argument of internal plastification is independently supported by annealing experiments. Figure 7a shows the evolution of peak height after annealing of all samples until a constant value of peak height was obtained. As one can see from this figure, the organization of all copolymers is considerably increased by annealing and the sharp order enhancement in BMA and OMA copolymers given in Figure 4 disappears as a result of annealing. This indicates that the order enhancement effect observed in un-annealed copolymers of ChMA and BMA or ChMA and OMA is due to an increase of segmental mobility. The increase of segmental mobility through annealing enhances the ordering of copolymers with high Tg and decreases the amount of disorder locked in by rapid evaporation of the solvent. It is interesting to note that the rate of the collapse of order as a function of the mole percent of the non-mesogenic component in the copolymer Figure 7b is much slower for MMA than for monomers with longer alkyl chains. A drastic increase in this rate takes place at $n > 3$ presumably indicating severe steric interference with the ordering of adjacent cholesteric side groups for comonomers such as BMA or OMA.

It is thus becoming increasingly apparent that mesomorphic order in macromolecules is strongly conditioned by the level of segmental motion in the polymer. High segmental mobilities "unlock" the amorphous structure of freshly precipitated polymers with mesogenic side groups and favor the formation of liquid crystalline order in such polymers. This can be accomplished either by internal plastification (lowering Tg) or annealing.

Acknowledgement

Acknowledgement is made to the donors of Petroleum Research Fund administered by the American Chemical Society for partial support of this research under the PRF Research Grant No. 8285AC6C. Thanks are expressed to the NSF for partial support under the Grant DMR-75-17397.

Literature Cited

1. Blumstein, A., Blumstein, R.B., Clough, S.B., Hsu, E.C., Macromolecules, (1975) 8, 73.

2. Blumstein, A., Clough, S.B., Patel, L., Blumstein, R.B., and Hsu, E.C., Macromolecules, (1976) 9, 243

3. Clough, S.B., Blumstein, A., and Hsu, E.C., Macromolecules (1976) 8, 123.
4. Hsu, E.C., and Blumstein, A., J. Polymer Sci. (letters), (1977) 15, 129.
5. Newman, B.A., Frosini, V., and Maganini, P.L., J. Polymer Sci., Polymer Phys. Ed., (1975) 13, 87.
6. Perplies, E., Ringsdorf, H., and Wendorff, J.M., Makromol. Chem., (1974) 175, 553.
7. Hardy, Gy., Nyitrai, K., and Cser, F., IUPAC Internat. Symp. Macromol. Chem., Budapest, (1969) 121.
8. Toth, W.J., and Tobolsky, A.V., J. Polymer Sci., (1970) B8, 289.
9. DeVisser, A.C., Feyen, J., deGroot, K., and Banties, A., J. Polymer Sci., (1972) B, 10, 851.
10. Tanaka, Y., Kabaya, S., Shimura, Y., Okada, A., Kurihata, Y., and Sakakibara, Y., J. Polymer Sci., (1972) B10, 261.
11. Saeki, H., Iimura, K., and Takeda, M., Polymer J., (1972) 3, 414.
12. DeVisser, A.C., deGroot, K., Teyen, J., and Bantjes, A., Polymer Letters, (1972) 10, 851.
13. Shibaev, V.P., Freidzon, J.S., and Plate, N.A., Dokl. Acad. Nauk. USSR, (1976) 227, 1412.
14. Samulski, E.T., and Tobolsky, A.V., Macromolecules (1968) 1, 555.
15. Tanaka, T., Shioraki, H., and Shimura, J., CA78 136543 (1973).
16. deVries, A., Molecular Cryst. Liquid Cryst., (1970) (1970) 10, 219.
17. Hsu, E.C., Clough, S.B., Blumstein, A., J. Polymer Sci. (letters) (1977) 15, 545.
18. Osada, Y., and Blumstein, A., J. Polymer Sci., (letters), (in print).

RECEIVED December 8, 1977.

6

Influence of Mesomorphic Order on the Physical Properties of Poly(*p*-Biphenyl Acrylate) and Related Polymers

B. A. NEWMAN

Department of Mechanics and Materials Science, Rutgers, The State University, New Brunswick, NJ 08903

V. FROSINI and P. L. MAGAGNINI

Università di Pisa, Istituto di Chimica Industriale ed Applicata, 56100 Pisa, Italy

In previous investigations (1,2) the influence of the chemical structure on the physical properties of aromatic polyacrylates and polymethacrylates, obtained by radical polymerization, has been extensively studied. The most interesting and stimulating result (3-6) was the discovery that some of these polymers, namely, poly(p-biphenyl acrylate) (PPBA), poly(p-cyclohexyl phenyl acrylate (PPCPA) and poly(p-acryloyloxyazobenzene) (PPAAB), display thermodynamic properties typical of crystalline polymers. Recently the crystalline character of atactic PPBA and PPCPA has been pointed out by the study of their dynamic mechanical behavior (7). However x-ray studies (8,9) have shown that the structure of these polymers is not crystalline in the conventional sense. A one-dimensional ordering showing in some cases as many as four sharp reflections and a periodicity of 23.2 Å was observed in a direction perpendicular to the main chain, together with a diffuse halo in the wide angle region. In order to account for these results a smectic structure was proposed (8), the long rigid side groups being randomly directed at right angles on both sides of the main chain to form a layered structure.

On the basis of this model it was anticipated (8) that a syndiotactic configuration of the macromolecules would have favored the attainment of a higher degree of order and perhaps also the formation of conventional crystalline structures. Studies have therefore been undertaken in order to synthesize samples of PPBA having different types of stereoregularity and to determine their morphology and physical properties.

Experimental

Materials. Atactic PPBA was prepared by radical polymerization in bulk or in benzene solution at 60-75°C, using Bz_2O_2 as initiator. It contained ∿55% syndiotactic diads. Isotactic PPBA was prepared by polymerizing the monomer in toluene at -78°C with Bu Li as initiator. The procedure was very similar to that reported in the literature for the preparation of other isotactic

0-8412-0419-5/78/47-074-071$05.00/0

polyacrylates. The degree of isotacticity was ∿90%. Fractionation with appropriate solvents led to a sample with a degree of isotacticity higher than 97%.

Monomers with bulky side groups do not form syndiotactic-rich polymers readily. Several attempts to prepare syndiotactic PPBA were made. The best results were achieved by polymerizing the monomer in toluene solution at -78° under U.V. irradiation, when a sample was obtained with ∿65% syndiotactic diads.

Atactic poly(p-biphenyl methacrylate) PPBMA was prepared by radical polymerization in bulk at 75°C. Isotactic and syndiotactic PPBMA were obtained by polymerization with Bu Li as initiator at -78°C in toluene and THF respectively.

Techniques. N.M.R. spectra of atactic, isotactic and syndiotactic samples were recorded using a Jeol PS-100 instrument and a Varian 220 spectrometer, equipped with variable temperature controller and frequency meter (10). Spectra were obtained using 5-10% (W/V) solutions in CCl_4 or O-chlorobromo-benzene with HMDS as internal standard.

Specific heat measurements were made using a Perkin-Elmer DSC-IB differential scanning calorimeter. The sample size was generally ∿15 mg. and the data was obtained for a sample surrounded by a dry nitrogen atmosphere. Synthetic sapphire was used as a reference. In each run, carried out at a rate of 8°/min, a temperature range of 20°K was explored. D.S.C. scans were obtained at scanning rates of 32°C/min.

The thermal optical data were obtained using a photocell in conjunction with a Bausch-Lomb 500μV VOM7 recorder, programmed by a Mettler hot stage.

X-ray diffraction data were obtained using a Norelco wide-angle diffractometer with Cu Kα radiation. X-ray diffraction patterns were also recorded on film using a flat plate, with a specimen to film distance of 7.0 cm. Cu Kα radiation was used so that the 23.2 Å spacing appeared on the same film with the wide-angle diffraction.

A thermogravimetric analysis was carried out for all polymers using a Mettler thermogravimetric analysis device. Both the absolute weight and the rate of weight loss were recorded as a function of temperature.

Results and Discussion

a. Nuclear Magnetic Resonance. The tacticities of all the polymers studied were determined from N.M.R. measurements using the method described in a previous (10) publication. The method is based on the consideration that the two backbone methylene protons of isotactic sequencies, because of their different shielding, are magnetic non-equivalent, while those of syndiotactic sequences, experiencing the same magnetic environment, are equivalent. For isotactic PPBA three bands with 1:1:1 relative intensities are expected, whereas syndiotactic PPBA should give two groups of lines with 1:2 relative intensities. The percent tacticity can be derived from the spectra using the relative areas under corresponding peaks. These results for all polymers investigated in this study are shown in Table I.

TABLE I

Polymer		Tacticity
Atactic PPBA		55% syndiotactic
Syndiotactic PPBA		65% syndiotactic
Isotactic PPBA	(a)	97% isotactic
	(b)	90% isotactic
	(c)	78% isotactic
Atactic PPBMA		80% syndiotactic
Syndiotactic PPBMA		90% syndiotactic
Isotactic PPBMA		90% isotactic

b. Differential Scanning Calorimetry. D.S.C. scans from isotactic, atactic and syndiotactic PPBA were made and the specific heats at various temperatures calculated. This data is shown in Figure 1 for both atactic and isotactic samples. The tacticity of the isotactic sample used for this figure was 97%. It can be seen that although the glass transition occurs at the same temperature for both samples, the isotactic sample melts at a temperature approximately 40°K lower than the atactic sample.

It was found that the melting point of atactic, isotactic and syndiotactic polymer varied with tacticity and thermal history. In order to investigate further these effects, polymer samples were crystallized isothermally at various temperatures and the melting point of each sample established using the D.S.C. scans. The results are summarized in Figure 2. The data shown here for the isotactic sample were obtained from a sample with a measured tacticity of 90%. It was found that the melting temperature increased linearly with crystallization temperature but that the rate of increase depended on the polymer configuration as shown in Figure 2. The rate of increase of melting temperature with crystallization temperature was considerably smaller for the atactic polymer than for the isotactic and syndiotactic polymers. An equilibrium melting temperature can be determined by extrapolation of the measured melting points as shown in Figure 2. The values obtained were 511°K for the isotactic polymer (90% isotactic), 550°K for the atactic polymer and 578°K for the syndiotactic sample.

As we have already discussed, the extent of syndiotacticity in the polymer here described as syndiotactic was actually quite low (($\sim$65%) Table I), and in fact only slightly higher than the polymer obtained by radical polymerization and here described as atactic (55% syndiotactic diads). From Figure 2 it can be seen

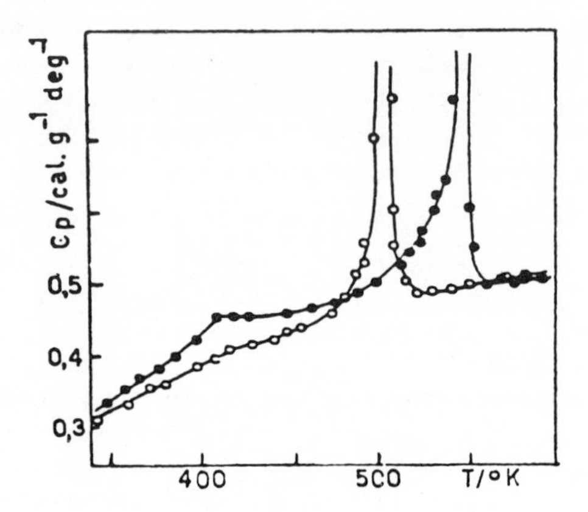

Figure 1. Specific heat data for atactic (●) and isotactic (○) PPBA

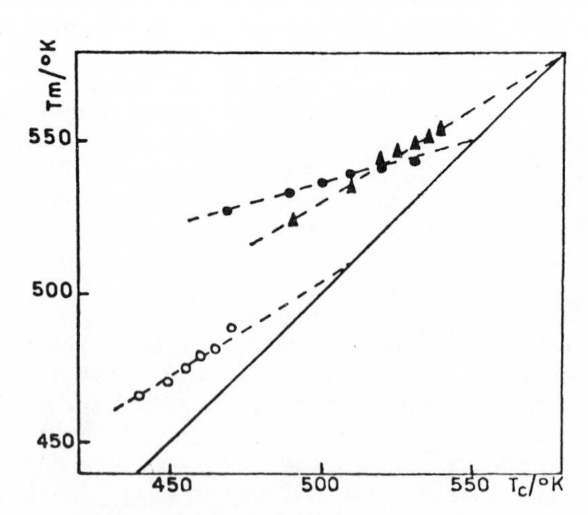

Figure 2. Variation of melting temperature with crystallization temperature for atactic (●), isotactic (○), and syndiotactic (△) PPBA

that the melting point of the atactic polymer may be greater than the melting point of the syndiotactic polymer at low crystallization temperatures. However the extrapolated melting temperature for the syndiotactic polymer is 28°K higher than for the atactic polymer. These results indicate that the equilibrium melting point of PPBA increases with increasing syndiotacticity. This supports the conclusion that a syndiotactic configuration of the macromolecules of PPBA favors the regular packing of the aromatic side groups, according to the suggested model.

The thermal properties of isotactic, atactic and syndiotactic PPBMA are illustrated by the D.S.C. traces in Figure 3. None of these samples showed a melting endotherm. Thus even a syndiotactic configuration (extent of tacticity 90%) is not able to promote efficient packing of the side groups for these polymers.

Interesting peculiarities were noted in the melting behavior of isotactic PPBA with a lower degree of isotacticity. Figure 4 shows the D.S.C. traces for a sample 78% isotactic. The most interesting observation is that for this polymer two widely separated endothermic peaks can be observed, at 490°K and 540°K. On the corresponding cooling curve two exothermic peaks at 500°K and 460°K were observed. Upon reheating the endothermic peaks were again observed at somewhat lower temperatures but the intensity of the high temperature peak was found to be greatly reduced.

The higher endothermic peak, for the isotactic sample, occurs at a temperature close to the melting point of the atactic polymer, and would appear, at first sight, to be associated with the fusion of atactic polymer present in the isotactic sample. This explanation was tested by the study of blends of isotactic polymer (97% isotactic) and atactic polymer. The D.S.C. traces did show two endothermic peaks at ∿490°K and ∿540°K. However, upon cooling and reheating, both high and low temperature peaks appeared again without any appreciable change. The explanation that the isotactic sample with a smaller extent of tacticity can be simply regarded as a blend of isotactic and atactic polymer is therefore not sufficient for a complete explanation.

c. Thermal Optical Analysis. A thin film of the polymer, cast from solution and observed between crossed polars in the polarizing microscope, showed uniform extinction and indicated that polymer precipitated from solution was essentially amorphous. However for the case of atactic PPBA, upon heating, some birefringence developed and at 200°C extensive regions of structure were observed. Further heating of the film led to a melting with a consequent decrease in birefringence to uniform extinction. Upon cooling from the melt, structure with birefringence again appeared.

A quantitative measure of this behavior was provided by replacing one the eyepieces with a photocell and attaching the photocell to a recorder to provide a measure of the light reaching the eyepiece. The heating and cooling was programmed using

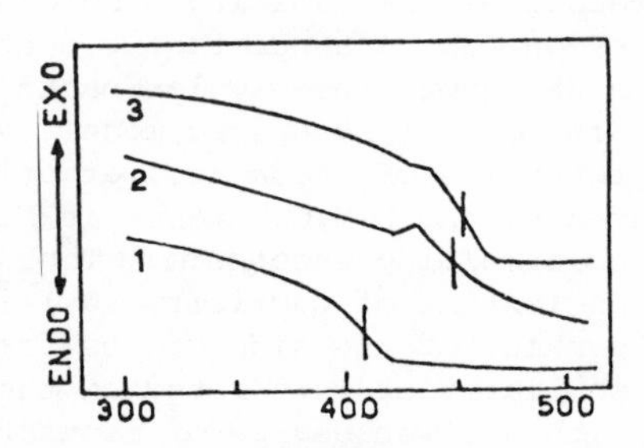

Figure 3. D.S.C. traces for isotactic (1), atactic (2), and syndiotactic (3) PPBMA

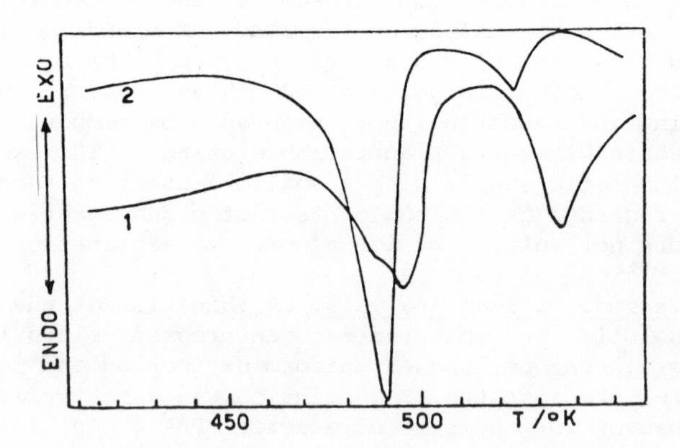

Figure 4. D.S.C. traces for isotactic PPBA sample (c); on first cooling (1) and second cooling (2)

a Mettler hot stage. Films were cast from chloroform solution directly on the microscope slide. The thickness of the film could be varied by changing the amount of polymer used. Usually a thickness of ∿0.5 mil was sufficient to give changes in birefringence of sufficient magnitude. The samples were heated and cooled at 3°C/min.

The results for atactic and isotactic PPBA are shown in Figure 5. The isotactic sample shown here was 78% isotactic. For both atactic and isotactic sample the initial cast film shows essentially zero birefringence. When the temperature exceeds ∿388°K (the glass transition temperature 385°K) the birefringence starts to increase, and reaches a maximum at ∿532°K for the atactic sample. Premelting and melting then reduce the birefringence until a zero value is reached at ∿543°K. The isotactic polymer however, with 78% tacticity, showed an increase in birefringence up to 493°K. After a small decline, a subsequent increase to a second maximum at 518°K was observed. Further increase in temperature led to a melting which was complete by ∿543°K. The highest birefringence observed was in fact only first order grey, so that the occurrence of the second maximum was quite unexpected.

Isotactic and atactic polymers both displayed the same behavior during cooling and reheating. Upon cooling from the melt a birefringent structure was obtained which upon subsequent reheating melted. For the atactic polymer zero birefringence was obtained for temperature ∿543°K. The isotactic sample did not show a second maximum in this case; the melting appeared to occur at the location of the first maximum, zero birefringence being observed at ∿523°K. It is interesting to compare the T.O.A. data for atactic PPBA in Figure 4 with the data obtained from isotactic polystyrene films cast in the same manner. This data is shown in Figure 6. Despite the different molecular configuration the optical behavior described here is essentially the same.

We assume that birefringence can arise from two sources; a crystallization (or ordering phenomenon) from the amorphous phase to give birefringent structures; and an orienting of macromolecular chains either in the crystalline or the amorphous phases. The clarification of these effects is best achieved using x-ray diffraction methods, and these results will presently be described. However a correlation of the T.O.A. data with the D.S.C. data can be made. It appears that the original film cast from solution is amorphous and without extensive molecular orientation. At temperatures above Tg the molecular mobility is sufficiently great to induce "cold" crystallization.

The T.O.A. data for isotactic PPBA parallel the D.S.C. data. If changes of birefringence were to be associated only with the melting process then a plateau in the T.O.A. data should have been observed rather than a maximum. This indicates that important textural changes must also accompany the complex melting behavior displayed by this isotactic polymer.

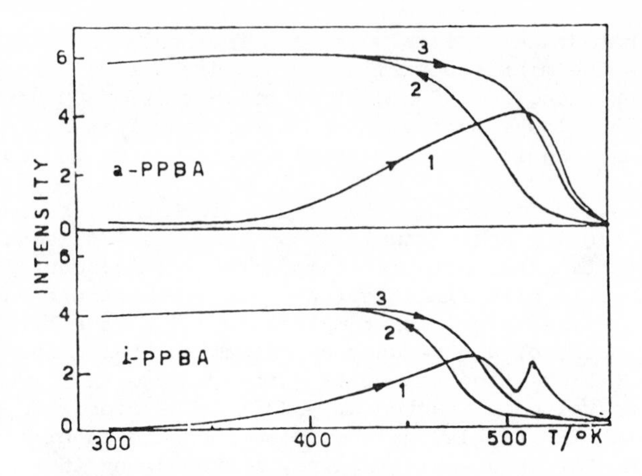

Figure 5. T.O.A. data for atactic and isotactic PPBA, on heating cast film (1), on solidification (2), and on reheating (3)

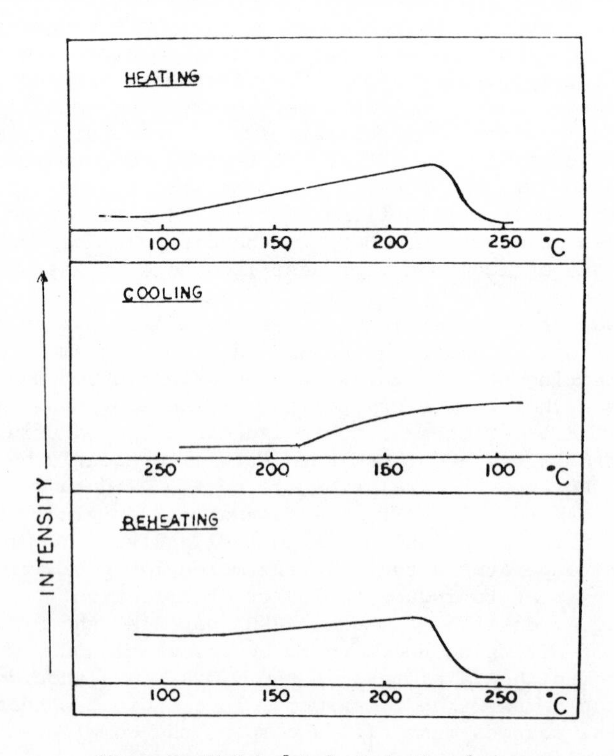

Figure 6. T.O.A. data for isotactic polystyrene

X-Ray Diffraction. X-ray diffractometer scans of various atactic PPBA samples are shown in Figure 7. Sample A was obtained by precipitation from $CHCl_3$ solution directly after polymerization. Samples B and C were obtained by casting thin films from $CHCl_3$ solution onto clean flat substrates (lead and glass). Sample D was obtained by slow cooling and solidification of molten polymer on a glass substrate to give a thin film. The preparation of sample C was essentially the same as that used to prepare films for the T.O.A. studies.

Sample D gave the diffraction pattern discussed in presious publications (9). It is characterized by a sharp reflection corresponding to a spacing of 23.2 Å and a broad halo at higher Bragg angles with a maximum intensity at 19.5°. Very weak 2nd and 3rd orders are present. Samples A and D show the same general characteristics. The low angle peaks are not as intense and careful measurement shows that these correspond to a spacing of 21.0 Å. Sample C showed very little diffraction at all except for a very small peak at the Bragg angle 19.5°.

Some annealing treatments were then carried out on samples A, B and C at various temperatures and times. These results are tabulated in Table II and the diffraction patterns from samples B and C are shown in Figures 8 and 9.

TABLE II

Sample	Annealing Treatment		Ratio I_s/I_D
	T°C	Time(hrs)	
A		none	1:4
	170	30 mins	1:1
	170	3 hrs	2:1
	210	3 hrs	3:1
	240	3 hrs	4:1
B		none	1:3
	170	30 mins	1:1
	170	3 hrs	2:1
	210	3 hrs	3:1
	240	3 hrs	4:1
C	170	30 mins	30:1
	170	3 hrs	50:1
	210	3 hrs	120:1
	240	3 hrs	140:1

After annealing, the low angle peak shifted to 3.8° (corresponding to an interplanar spacing of 23.2 Å), became more sharp, and increased in intensity relative to the wide-angle halo. A measure of this increase in intensity is given in Table II where the

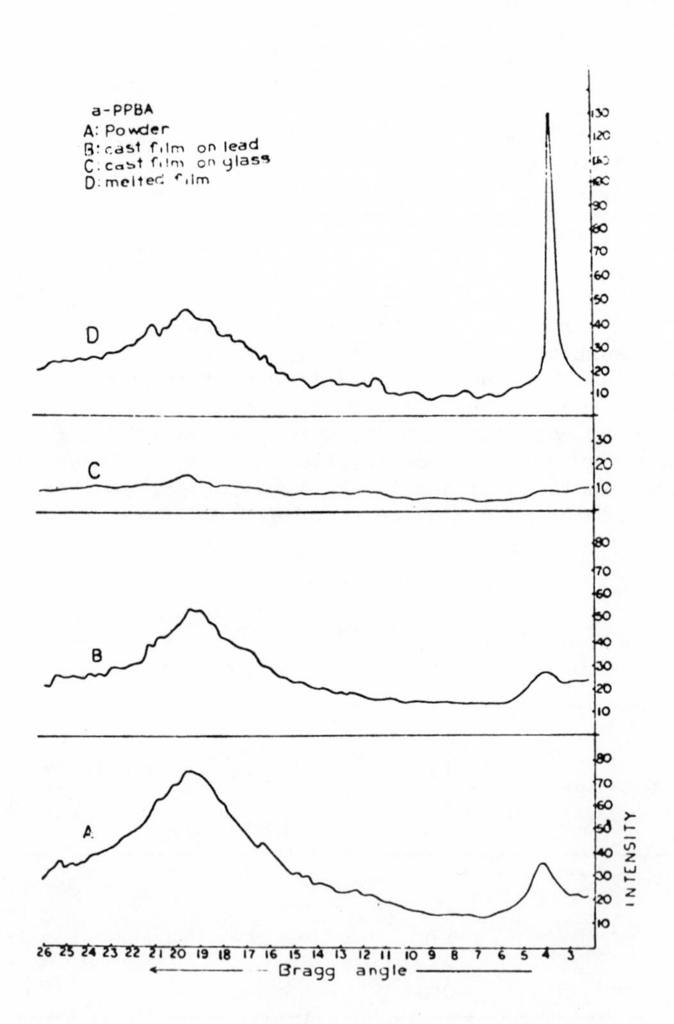

Figure 7. X-ray diffraction from atactic PPBA

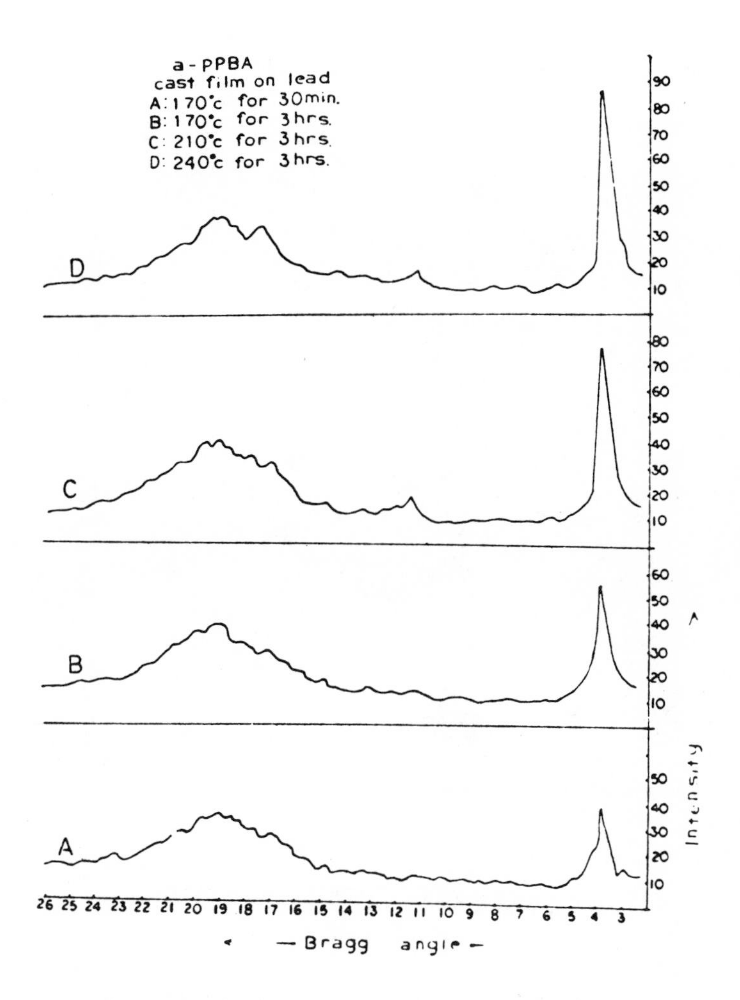

Figure 8. X-ray diffraction scans from annealed films cast on lead

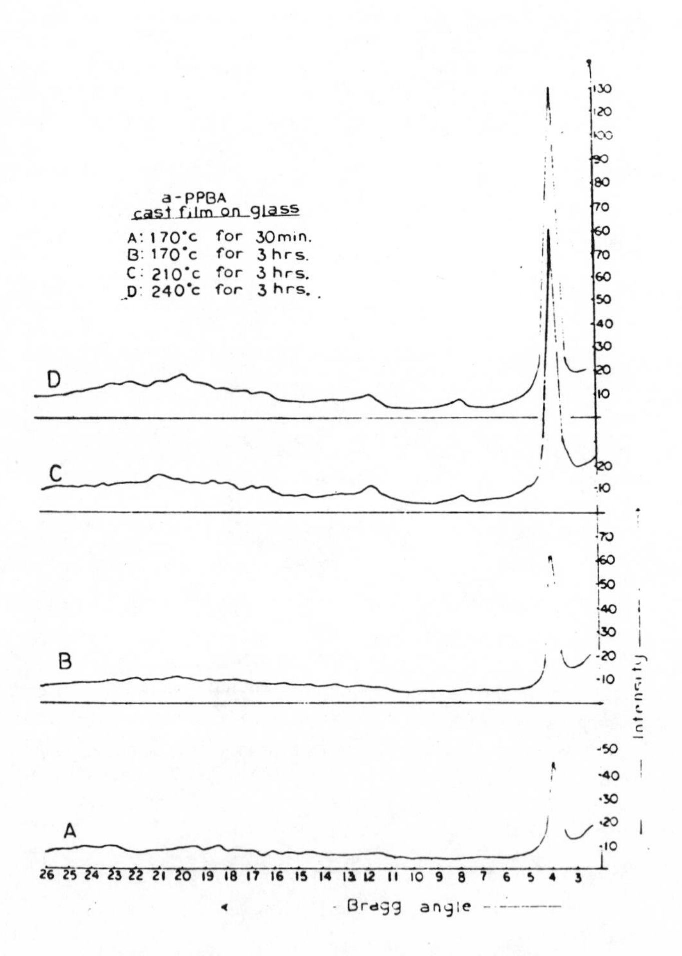

Figure 9. X-ray diffraction scans from annealed films cast on glass

intensity ratio (I_S/I_D) of the low angle peaks height (I_S) to the wide-angle peak height (I_D) is listed. The most noticeable feature was that for sample C this ratio increased to only 4:1. Moreover, for sample B the annealing treatment led to the appearance of a small peak at 2θ = 17.5° superimposed on the broad halo.

In all respects samples A and B behaved the same. It is likely that the film cast lead is very similar to the random powder aggregate obtained by precipitation from solution. However the film cast on the glass surface shows very different x-ray diffraction data. After annealing, the low angle reflection shows a much greater intensity relative to the wide angle halo. This can be understood if there is some preferred orientation of the layered structure in these cast films, both the extent of ordering and the degree of preferred orientation increasing with annealing temperatures and times. If the smectic layered structure is oriented parallel to the glass substrate surface, such an enhancement of the low angle reflection would be anticipated for the reflection geometry utilized by the diffractometer.

Studies of the isotactic PPBA were also carried out. The cast films showed an initial small low angle peak corresponding to an interplanar spacing of 23.2 Å rather than 21.0 Å observed for the case of the atactic polymer. In all other respects, including the preferred orientation, the x-ray diffraction data was very similar to the data obtained from the atactic polymer.

Thermogravimetric Analysis (T.G.A.)

For further characterization, degradation of the polymer was studied using the Mettler thermogravimetric apparatus. Figure 10 shows the weight loss curves for atactic PPBA when heated at 6°C/min. No appreciable weight loss occurred at temperatures up to 613°K. Isotactic polymers showed similar traces and no appreciable degradation occurred at temperatures less than 573°K. These data confirm that degradation processes do not appear to play a role in the thermal studies described here.

Conclusions

The results described here indicate that differences in the degree and type of tacticity of PPBA samples fail to influence the structure of these polymers as revealed by x-ray diffraction. This is a very important finding since it implies that the structure of the polymer is, in this case, independent of the polymerization procedure. Several authors have implicitly granted the possibility to obtain polymers with mesomorphic structure lies in the availability of mesogenic monomers, and on the possibility to polymerize them in the mesomorphic phase. Our results contradict the suggestions. The only difference one can reasonably expect between polymers prepared in the mesomorphic and isotropic phase

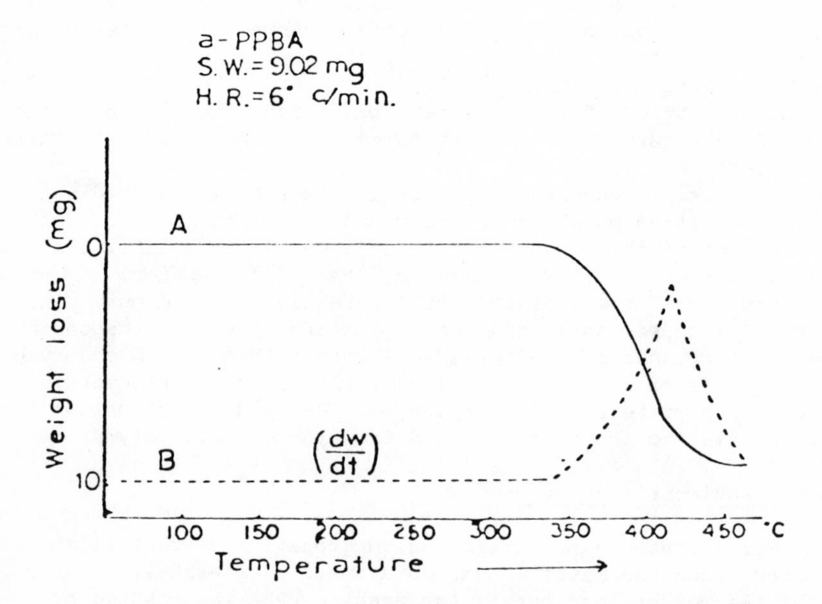

Figure 10. T.O.A. data for atactic PPBA

(besides degree of polymerization) may be in relation to a different stereoregularity. Our results show that even marked differences in stereoregularity do not alter the smectic structure of PPBA, thus demonstrating that the lateral ordering depends critically on the very structure of the aromatic side groups rather than the main chain configuration.

Our results also indicate that thermal history and sample preparation conditions can alter the morphology and texture of the polymer and also influence the melting point. The preferred orientation of the smectic layered structure of the cast films on the glass substrate is not altogether unexpected since such observations are commonly found among mesogenic compounds.

The different thermal behavior of PPBA samples with different tacticities is intriguing and perhaps associated with small differences in the mode of packing of aromatic side groups within the layers, not observable by x-ray diffraction.

Our results also indicate that the melting point of PPBA increases with increasing extent of syndiotacticity. This supports the conclusion that a syndiotactic configuration favors the model previously proposed for the layered smectic structure.

Literature Cited

(1) Pizzivani, G., Magagnini, P. L., and Giusti, P., J. Polym. Sci. (1971) 9, A2, 1133.

(2) Pasquini, M and Frosini, V., IUPAC International Symposium on Macromolecules, Madrid 1974, Reprint III 5-13.

(3) Baccaredda, M, Magagnini, P. L., Pizzivani G., and Giusti, P., Polym. Letters (1971) 9, 303.

(4) Magagnini P. L., Marchetti, A., Matera, F., Pizzivani, G., and Turchi, G., European Polym. J., (1974) 10, 585.

(5) Frosini, V., Magagnini, P. L., and Newman, B. A., J. Polym. Sci. (1974) 12, 23.

(6) Balloni, P., Thesis, University of Pisa (1972).

(7) Frosini, V., Magagnini, P. L., and Newman, B. A., J. Polym. Sci. in press.

(8) Newman, B. A., Magagnini, P. L., and Frosini, V., Advances in Polymer Science and Engineering, Pae, K. D., Morrow, D. R., and Chen, Yu, Plenum Press, New York (1972) 21.

(9) Newman, B. A., Frosini, V., and Magagnini, P. L., J. Polym. Sci. Phys. Ed. (1975) 13, 87.

(10) Ceccarelli, A., Frosini, V., Magagnini, P. L., and Newman, B. A., Polym. Letters (1975) 13, 101.

RECEIVED December 8, 1977.

7

A New Series of Mesomorphic Monomers: Bulk Polymerization of 4-Cyano-4′-biphenyl (n + 2)-alkenoates

J. C. DUBOIS, J. C. LAVENU, and A. ZANN

Thomson-CSF, Domaine de Corbeville, 91401 Orsay, France

Polymers which exhibit mesomorphic order may have interesting anisotropic properties such as high optical, electrical, or mechanical anisotropies. Such polymers may be obtained either by polymerization of a non-mesomorphic monomer in a liquid crystalline solvent (1,2) or by direct polymerization of a mesomorphic monomer in its liquid crystalline state (3-9). The organization in the resulting polymers may be improved if the polymerization is done under orienting conditions in electric or magnetic fields or within suitable boundary conditions (10,11,12,13).

In this chapter a new series of mesomorphic monomers is presented -- the 4-cyano-4'-biphenyl (n+2)-alkenoates. Their bulk polymerization in the nematic state has been investigated, and thin layers obtained by thermal polymerization with or without an orienting electric field have been studied. Dielectric properties are discussed also.

Monomers

The monomers which have been synthesized have the general formula:

$$CH_2 = CH - (CH_2)_n - COO - \langle O \rangle - \langle O \rangle - CN$$

with n equal to 0, 1, or 2.

Synthesis. The preparation of these compounds involved three steps. The first step was the synthesis of 4-cyano-4'-hydroxybiphenyl (14):

$$HO - \langle O \rangle - \langle O \rangle - Br \xrightarrow[\text{N-methyl-2-pyrrolidone (reflux)}]{\text{Cu CN - 90 min}} HO - \langle O \rangle - \langle O \rangle - CN + CuBr$$

0-8412-0419-5/78/47-074-086$05.00/0

The complex was broken down by using concentrated hydrochloric acid and ferric chloride in aqueous solution at 60°C. The second step was the preparation of the acid chloride from the selected vinylic acid (15):

$$CH_2{=}CH\text{-}(CH_2)_{\underline{n}}\text{-}COOH + \langle O \rangle\text{-}COCl \xrightarrow[\text{heating}]{} CH_2{=}CH\text{-}(CH_2)_{\underline{n}}\text{-}COCl$$

The acid chloride was simultaneously distilled and collected on cool hydroquinone. The last step, esterification, was conducted at room temperature in N,N'-dimethylaniline medium (16):

$$CH_2{=}CH\text{-}(CH_2)_{\underline{n}}\text{-}COCl + HO\text{-}\langle O \rangle\text{-}\langle O \rangle\text{-}CN \xrightarrow{48\ hr} CH_2{=}CH(CH_2)_{\underline{n}}COO\text{-}\langle O \rangle\text{-}\langle O \rangle\text{-}CN$$

The resulting compound was purified by chromatography on a silica gel column with a benzene/hexane mixture as eluant and then was recrystallized two or three times in ethanol.

Mesomorphic Properties. The synthesized compounds have been investigated by means of a polarizing microscope equipped with a heating and cooling stage (Mettler FP5). The temperature and enthalpies of transition were determined by differential scanning calorimetry (DSC 1 B-Perkin Elmer). The results are listed in Table I where K, N, I are, respectively, the crystalline, nematic, and isotropic phases, and ΔH_m is the melting enthalpy. The mesomorphic properties differ greatly according to the length of the chain. For comparison, the compounds of the homologous saturated series (17) are mesomorphic at lower temperatures. The presence of the vinyl linkage in the unsaturated compounds stabilizes the mesophase.

Table I. Transition Temperatures and Enthalpies for the 4-Cyano-4'-biphenyl(n+2)-alkenoates

n	K		N		I	ΔH_m (Kcal/mol)
0	.	100	.	130	.	4,4
1	.	136	.	166	.	4,6
2	.	73	.	[73]	.	3,7

In order to widen the nematic range of the monomers and to decrease their melting temperatures, we have made binary mixtures of compounds n = 0 and n = 1 (Table I). We observed that the experimental diagram did not exhibit an eutectic, as is generally observed with other liquid crystals. The two compounds are probably miscible in all proportions in their crystalline state (solid solution). The phase diagrams, experimental and calculated, are plotted on Figure 1.

The dielectric anisotropy ϵ_a of the first monomer, the 4-cyano-4'-acryloyloxybiphenyl, was measured. The ϵ_a of a 0.10/0.90 molar-fraction mixture of this compound, with a nematic compound having an ϵ_a equal to 0.1, is 2.1. Since the anisotropies are additive, the anisotropy of the considered monomer is approximately equal to 2.0.

Bulk Polymerization

Bulk polymerization of each monomer has been studied by means of differential scanning calorimetry. The procedure for following the kinetics was similar to that described by Blumstein and Hsu (18). A small quantity of the compound was sealed in a DSC sample pan. The temperature was increased linearly to the melting point and was then set quickly to the desired value. The isothermal polymerization was performed during a given time t. At the end of this time, the sample pan was quenched with a dry ice-methanol mixture to stop the polymerization and to allow the recrystallization to take place. A thermogram was then taken again. For each polymerization time, a new sample was used. The rate of conversion in % is equal to $(1 - \Delta H/\Delta H_o) \times 100$ where H_o is the melting enthalpy of the pure monomer, and ΔH is the transition enthalpy of the remaining monomer after partial polymerization. Thermal polymerization of the first monomer of the series in its nematic state is very fast. The evolution of its melting peak is shown in Figure 2. This figure also shows that the presence of oxygen reduces the rate of polymerization. Since the two other monomers need addition of initiator to polymerize, only the kinetic studies of thermal polymerization of the 4-acryloyloxy-4'-cyanobiphenyl are presented. The rates of conversion in wt. % vs. time for three temperatures are plotted on Figure 3. The curves are perfectly linear up to 80% conversion, monomer and polymer are miscible, and no phase segregation occurs. From these curves, the logarithm of the polymerization rate vs. $1/T$ has been plotted (Figure 4), and the energy of activation E_a calculated. E_a in the nematic phase is equal to 119 k J/mol.

The bulk polymer obtained after 80 min at 110°C in N_2 atmosphere is white, very hard, and brittle. The texture of a thin film of this polymer looks like a disordered mesophase texture. The polymer is insoluble in all the classical solvents except in the dimethylformamide. It swells in chloroform. The thermogram of this polymer does not exhibit any transition from room temperature until 260 C.

Thin Layer Polymerization of 4-Acryloyloxy-4'-cyanobiphenyl

Investigation of optical and electrical properties of this polymer may be facilitated by making thin layers. Thus the polymerization has been done in a 50-μm thick cell. Such a cell

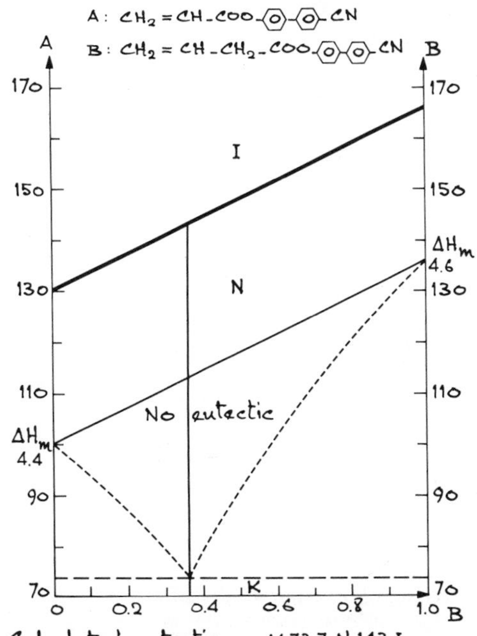

Figure 1. Phase diagram

Inhibitory affect of oxygen

atmosphere : nitrogen

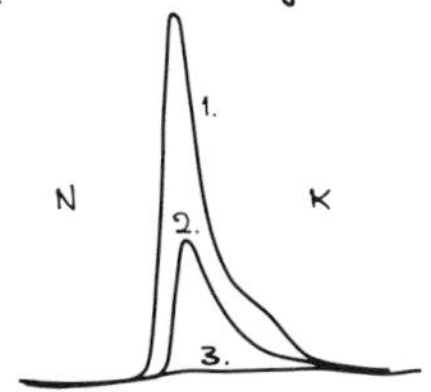

atmosphere : air

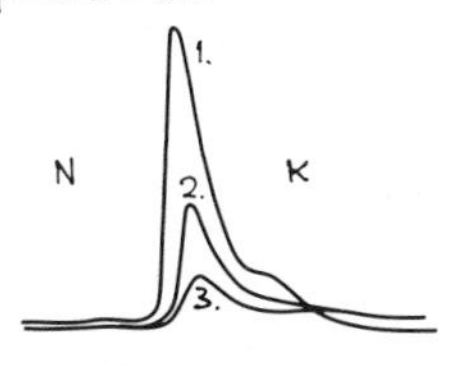

Evolution of the melting peak.

1. t = o
2. after 10 mn at 110°C
3. after 40 mn at 110°C

Figure 2. Polymerization of 4-cyano-4'-acryloyloxybiphenyl

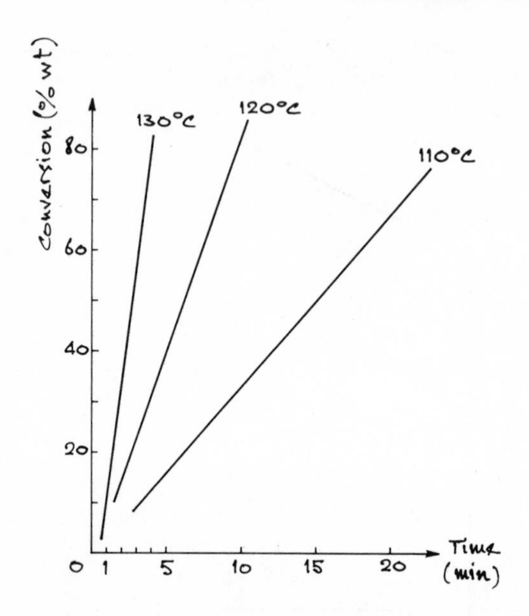

Figure 3. Conversion of 4-cyano-4'-acryloyloxybiphenyl to polymer as a function of time at various temperatures

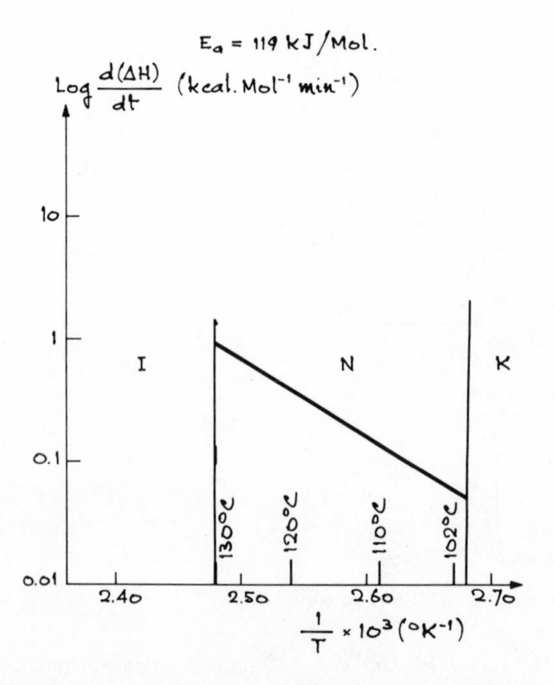

Figure 4. Arrhenius plot for the polymerization of 4-cyano-4'-acryloyloxybiphenyl in the nematic phase

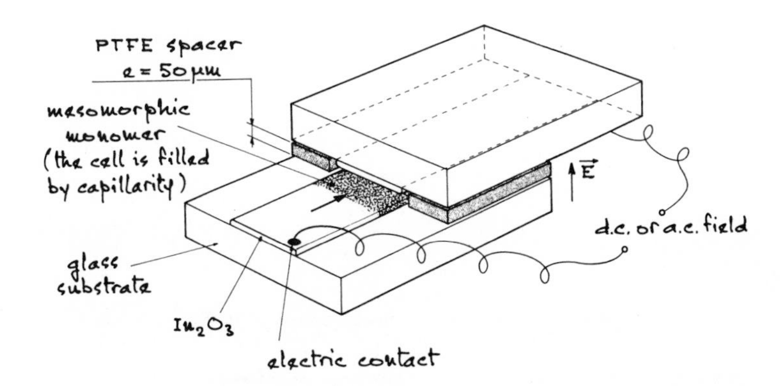

Figure 5. Cell for thermal polymerization of a 50-μm film of a liquid crystal monomer orientated by an electric field

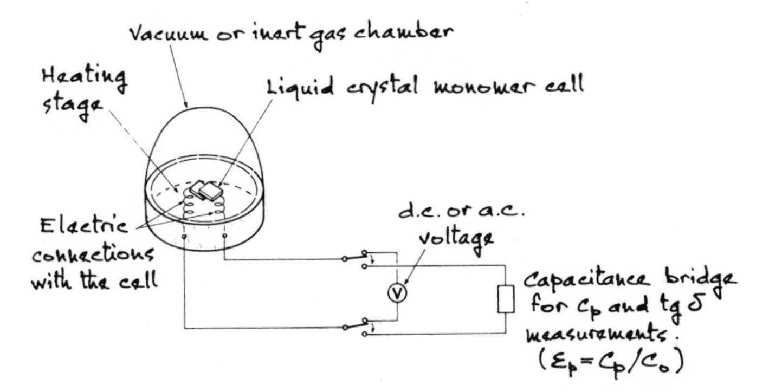

Figure 6. Apparatus for thermal polymerization of mesomorphic monomers under electric field

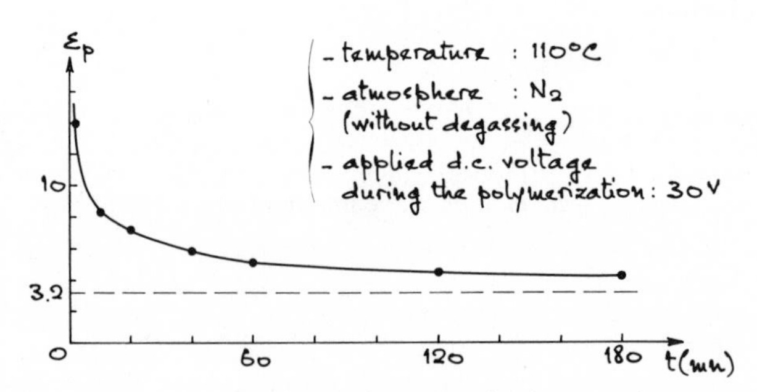

Figure 7. Evolution of the static dielectric constant ϵ_p of 50-μm film of $CH_2 = CH–COO–$⟨O⟩–⟨O⟩$–CN$ during polymerization

(Figure 5) is made up of two conductive glass plates separated by Teflon spacers 50 μm thick. The monomer in the solid state was placed at the inlet of the cell, degassed under 20 mm Hg at 70°C, and then introduced by capillarity on melting at 100°- 102°C. Polymerization was achieved at 110°C in 90 min under N_2 atmosphere. Well polymerized layers, displaying flat thermograms up to 260°C, were obtained. Observation with polarizing microscope shows birefringent and disordered texture.

A long-distance order is induced in a nematic liquid crystal by applying an appropriate magnetic or electric field. If such a field is applied during polymerization, a mesomorphic order may be at least partially maintained. Therefore we have performed a series of thin layer polymerizations under d.c. or a.c. electric field with values from 6 to 20 kV/cm. The capacity of the cell was measured simultaneously during the polymerization by an automatic capacitance bridge (H.P. 4270A). The scheme of the apparatus is drawn in Figure 6. An example of the evolution of the dielectric constant ϵ_p of the polymer being formed is given in Figure 7. The last ϵ_p value is independent of the value of the applied fields and always reaches 3.2. We can conclude that the polarization of orientation, which is very important in the monomer when $\epsilon_{//} \geq 20$, disappears in the polymer. The dipoles are blocked. In order to know if the mesogenic side groups have kept the mesomorphic, smectic, or nematic order, x-ray investigations are now being performed with a small-angle diffractometer. Comparison will be made with the results obtained on the poly-(p-biphenyl acrylate), which exhibits a smectic A order with spacing of about 20 Å between the smectic plane (19,20).

Conclusion

Three mesomorphic monomers belonging to a new series, the 4-cyano-4'-biphenyl (n+2)-alkenoates, were prepared. One of them, the 4-cyano-4'-acrylatebiphenyl, is nematic from 100 to 130° C; its thermal polymerization in the nematic phase is easy. A 50-μm polymerized film has been placed between two conductive glass plates. When the polymerization of such a layer is performed under orienting electric fields, the dielectric constant decreases from the high $\epsilon_{//}$ value of the liquid crystal to the classical value for the polymer. Mesomorphic order in the resulting polymer will be investigated by x-ray studies.

Literature Cited

1. Krentsel, B.A., Amerik, Y.B., Vysokomol Soedin. (1971) A 13 n° 6 1358.
2. Blumstein, A., Billard, J., Blumstein, R., Mol. Cryst. Liq. Cryst. (1974) 25, 83.
3. Paleos, C.M., Labes, M.M., Mol. Cryst. Liq. Cryst. (1970) 11, 385.

4. Liebert, L., Strzelecki, L., et al. Bull. Soc. Chim. Fr. (1973) n°2,597.
5. Ibid., (1973) n°2, 603.
6. Ibid., (1973) n°2, 605.
7. Ibid., (1975) n°9, 10, 2073.
8. Ibid., (1975) n°11-12, 2750.
9. Perplies, V.E., Ringsdorf, H., Wendorff, J.H., Ber. Bunsenges. Phys. Chem. (1974) 78, 921.
10. Liebert, L., Strzelecki, L., C.R. Acad. Sci. Paris (1973) 276, n° 8, 648.
11. Perplies, E., Ringsdorf, H., Wendorff, J.H., Polym. Lett. Ed. (1975) 13, 243.
12. Lorkowski, H.J., Reuther, F., Plaste Kautsch. (1976) 23 Jahrgang, 81.
13. Clough, S.B., Blumstein, A., Hsu, E.C., Macromolecules (1976) 9, 123.
14. Gray, G.W., Harrison, K.J., Nash, J.A., Constant, J., Hulme, D.S., Kirton, J., Raynes, E.P., Liq. Cryst., Ordered Fluids (1974) 2, 637.
15. Stempel, G.H., et al., J. Am. Chem. Soc. (1950) 72, 229.
16. Strzelecki, L., Liebert, L., Bull. Soc. Chim. Fr. (1973) 2, 597.
17. Dubois, J.C., Zann, A., J. Phys. Colloq. C3 (1976) 37, C3-35.
18. Hsu, E.C., Blumstein, A., Polym. Lett. Ed. (1977) 15, 129.
19. Newman, B.A., Frosini, V., Magagnini, P.L., ACS Symp. Ser. (1978) 74, 71.
20. Clough, S.B., Blumstein, A., de Vries, A., ACS Symp. Ser. (1978) 74, 1.

RECEIVED March 2, 1978.

8

Polymerization in the Liquid Crystalline State: Monomer-Polymer Interactions

F. CSER, K. NYITRAI, and G. HARDY

Research Institute for Plastics, H-1950 Budapest, Hungary

Sadron et al. (1) prepared lyotropic liquid crystalline systems using polymers and a polymerizable solvent and attempted to fix the mesomorphic structure by polymerizing the monomeric solvent. Bouligand et al. (2) attempted to prepare liquid crystalline substances with fixed structure by copolymerizing mono- and bifunctional mesomorphic monomers. However, neither group investigated the phase conditions of the monomers or of the monomer and polymer. In both cases identical homogeneous phases were assumed before and after polymerization.

In solid-state polymerization, however, the polymer formed in the reaction interacted with the initial monomer phase, forming a new thermodynamic system (3,4,5,6,7). The reactions were either heterogeneous or homogeneous phase topochemical reactions. The polymer remained isomorphous with its monomer forming a one-phase system in homogeneous reactions only (3,6). This type of reaction is, however, very rare (7,8,9,10). Heterogeneous reactions, in which the polymer and the monomer are not isomorphous, are much more frequent. In many cases, polymerizations starting as homogeneous change into a reaction which is predominantly heterogeneous. The determining condition for a homogeneous reaction is the isomorphism of the tactic polymer with the monomer crystals. This isomorphism can only be realized under special conditions required by the thermodynamics of the system. These conditions are as follows: the chain period of the polymer must coincide with a translational period of the monomer crystal lattice, the overlapping volumes of monomeric units in the polymer chain should not differ greatly from that of the monomer molecule (11), and the volume contraction during the chain formation should not be in the direction of the chain growth. These conditions are fulfilled in crystals of monomers with a long paraffinic chain substituent (12). Figure 1 shows an example of the isomorphous solid solution of poly(cetylvinylether) in its single layer of *pgg* layer symmetry. The isomorphism of monomer crystals and of the

0-8412-0419-5/78/47-074-095$05.00/0

polymer formed has been reported by several authors (8,9,10,13).

The topochemical aspect of the reaction varies depending on the type of mesomorphic state. Topochemical polymerization can be expected in smectic mesophases (12) where the functional groups of the layer formed by the parallel chains are located in one plane (14,15). Homogeneous topochemical reactions may proceed in a layer with either uni- or bimolecular structure. In the polymerization of cetylvinylether (16) the polymer formed remained isomorphous with the monomer, and the melting point of the system increased. This isomorphism was destroyed, however, when the system reached its melting point. After cooling it formed a two-phase system.

In the polymerization of cholesterylacrylate the isomorphism of the polymer with the monomer could be preserved up to the conversion limit of 70% at 0° C (17), but at 30° C the system became rapidly heterogeneous because of the increased mobility of the molecules. At both temperatures the monomer was in the smectic G state. The experimental evidence is shown in Figure 2.

The decrease in the melting and/or clearing points of the monomers caused by the presence of the polymer was detected in both the smectic state polymerization of vinyl oleate (18,19) and in the polymerization of cholesterylacrylate in the cholesteric state (17). The polyvinyl oleate formed in this reaction was crystalline while the polycholesterylacrylate was amorphous. The temperature of the phase transition was reduced in both cases.

The situation is completely different in the cholesteric and the nematic states. The rates of polymerizations are usually lower, and the activation energies of reactions are higher than those in the liquid isotropic state measured or extrapolated to the same temperature (20,21,22). The cholesteric and nematic states do not seem to favor topochemical processes, and, therefore, a structural isomorphism cannot be expected either. Because in this state the molecules are not packed densely, the formation of a homogeneous solution also can be expected with greater differences in overlapping volumes. The polymer/monomer state diagram of p-methyl, p'-acryloyloxyazoxybenzene (23) in Figure 3 shows that near the melting point of the monomer a heterogeneous two-phase system is formed in which both phases have a nematic structure. Above the melting point of the monomer the solution is a homogeneous nematic one; the polymer is dissolved in the isotropic liquid-phase monomer. Therefore, in the nematic phase of the monomer a heterogeneous phase polymerization can be expected. When the reaction is started in the isotropic liquid phase, it is homogeneous up to a given conversion, where a phase separation takes place. Thereafter, as the reaction proceeds, the nematic polymer/monomer solution dissolves the remaining isotropic

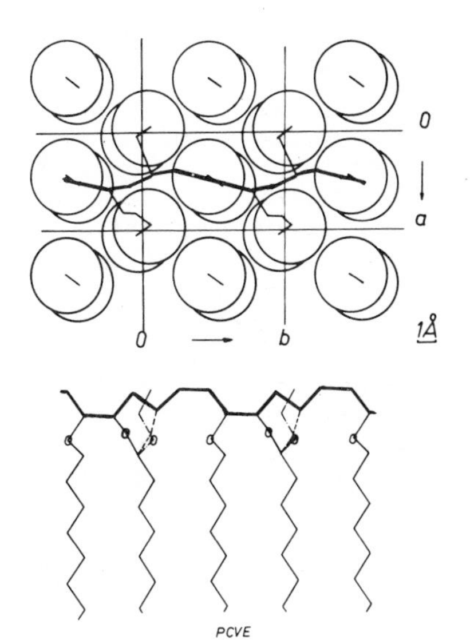

Figure 1. Polymer/monomer solid solution in cetylvinylether single layer of pgg *symmetry*

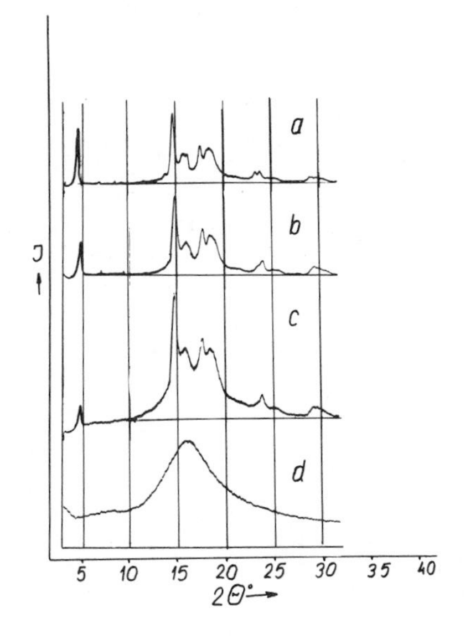

Figure 2. X-ray powder pattern of cholesterylacrylate monomer (a), of its precipitated polymer (d), of polymerizing systems polymerized at 0°C up to 70% conversion (b), and at 30°C up to 28% conversion (c).

monomer, and the reaction becomes homogeneous. The $\underline{T}g$ of the plasticized polymer system is reached, and the reaction terminates.

The polymer/monomer state diagram of cholesterylvinyl succinate shows another type of interaction (24,25) as can be seen on Figure 4. In one of the three possible crystalline phases a homogeneous topotactic solid state polymerization takes place. The polymorph is formed in the presence of even a trace of polymer, and there is a real isomorphism of the polymer and of the monomer. The highest polymer content in this solid solution is ca. 25%. At higher temperatures the isomorphous polymer/monomer system has a smectic structure. Thereafter, in the cholesteric state the system becomes heterogeneous. In samples with a high polymer content two smectic states were detected. A low viscosity smectic structure could be detected at higher temperatures.

The polymerization in the cholesteric state of this monomer is a heterogeneous phase reaction. The polymer formed is precipitated as a smectic phase, and as the conversion increases, the cholesteric monomer is dissolved in the isotropic system, and a homogeneous phase is formed. As the reaction proceeds further, the system becomes smectic (25).

For a long time we searched without success for a homogeneous reaction in the cholesteric state until we began to study the cholesterylvinylfumarate monomer. In the following we present the polymer/monomer state diagram of this substance.

Figure 5 shows the phase transition points found in this system with a 10% polymer content. Figure 6 displays the DSC transition heats as a function of the composition. The next two figures show the polymer/monomer state diagrams determined by DSC (Figure 7), by polarizing microscopy, and by thermomechanical methods (Figure 8). The figures show a good agreement between different methods. Figure 9 displays wide-angle x-ray diffractograms of some characteristic compositions of the polymer/monomer systems. The monomer is in the smectic G state (26). Systems with polymer content of less than 40% consist of two phases -- an amorphous phase of the plasticized monomer and a smectic B phase of the monomer containing the polymer. When the polymer content is greater than 40%, only the amorphous phase is detected. The upper part of Figure 9 shows the x-ray diffractograms of systems polymerizing at 145°C in the presence of 0.2% benzoylperoxide as a function of reaction time. In these polymerizing systems similar interactions occur as in the melted and cooled systems. Samples with polymer contents less than 50% display the green-blue reflection characteristic of the cholesteric state.

The characteristics of the states are as follows: When the polymer content is higher than 70%, the systems consist of a homogeneous phase which contains the plasticized

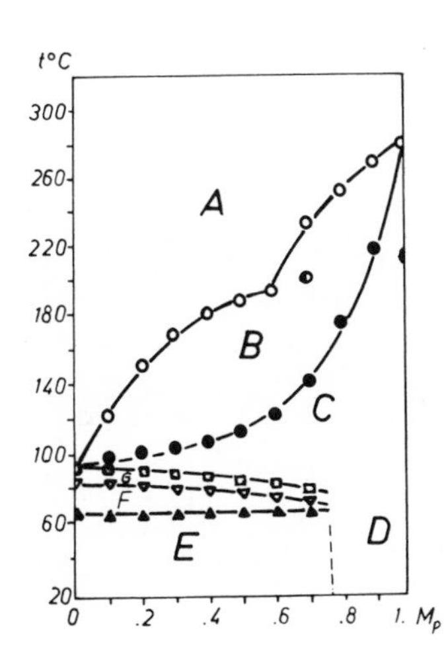

Figure 3. Polymer/monomer state diagram of p-methyl, p'-acryloyloxyazoxybenzene obtained by polarizing microscopy M_p *composition (weight fraction of polymer). Characteristic points: (▲), decreasing birefringence; (▼), mesomorphic transition; (□), mesomorphic melting; (●), solidus; (○), liquidus. The meaning of the areas: (A), isotropic liquid; (B), isotropic liquid + mesomorphic plasticized polymer; (C), liquid; (D), glassy states of plasticized polymer in mesomorphic phase; (E), crystalline monomer + mesomorphic plasticized polymer in the glassy state; (F), crystalline monomer + mesomorphic plasticized polymer in the liquid state; (G), mesomorphic monomer + mesomorphic plasticized polymer in the liquid state.*

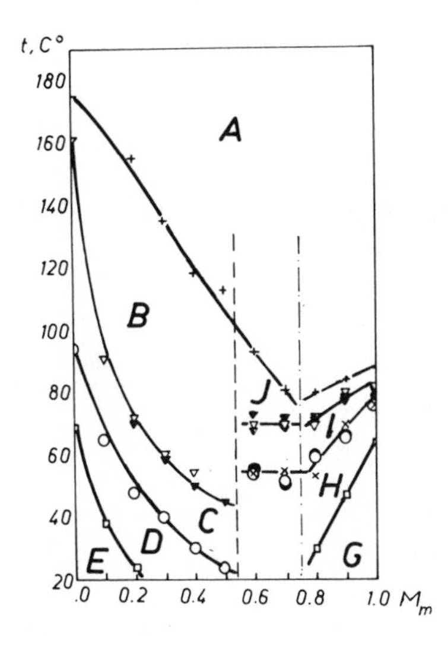

Figure 4. Polymer/monomer state diagram of cholesterylvinylsuccinate. M_m, *composition (weight fraction of monomer). (□),* T_g; *(○), change in the thermal exponent of deformation; (▽),* T_f, *all measured by thermomechanics. (▼), shift in the base line, if* $M_m < 0.5$ *or second peak if* $M_m > 0.5$; *(●), first peak on the DTA trace. (+), clearing point; (▽), melting begins; (×), phase transformation, all measured by polarizing microscopy. The meaning of the areas: (A), isotropic liquid; (B, C, D, and E), homogeneous mesomorphic plasticized polymer in liquid (B), in high elastic (C and D), and in glassy state (E), respectively. Solid solutions are in solid (G) and in plasticized states (H). Mesomorphic solutions are in smectic (I) and in cholesteric (J) states. G, H, I and J are biphasic in the concentration range of* $0.5 < M_m < 0.8$.

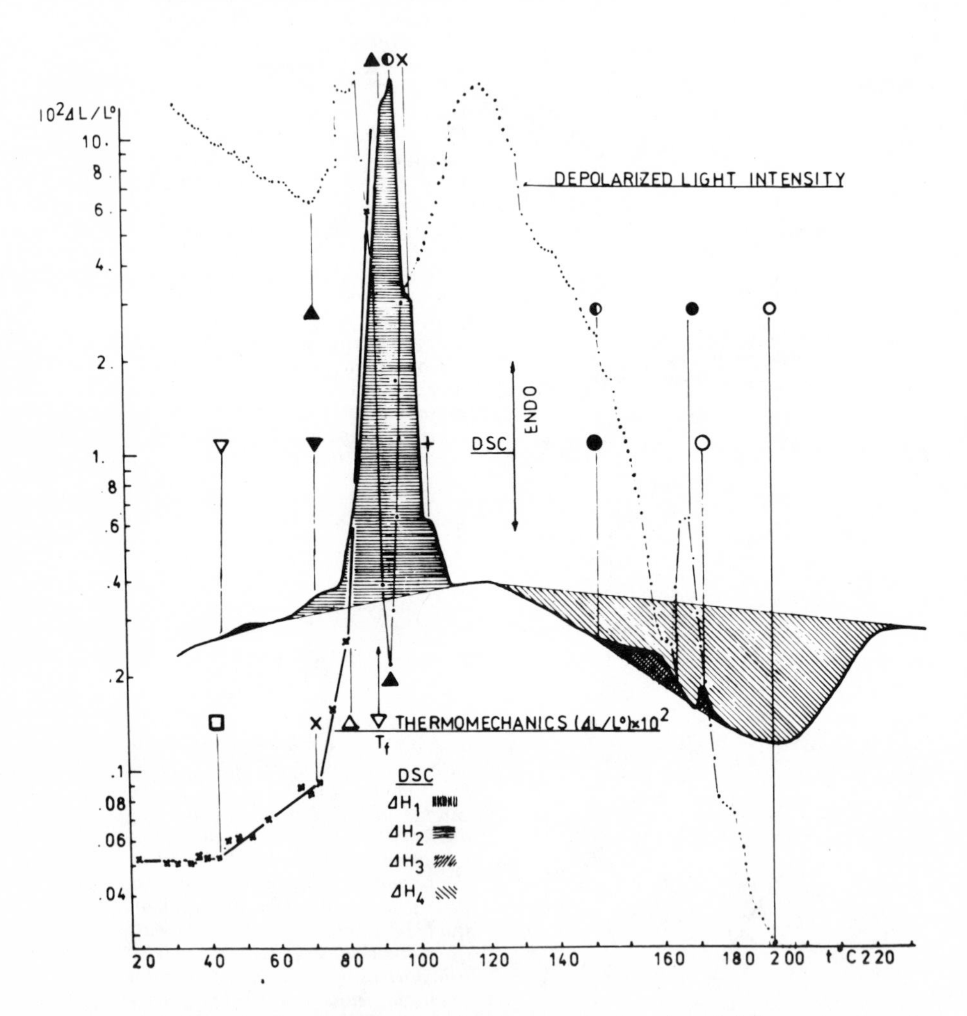

Figure 5. DSC traces (——), depolarized light intensity (· · ·), and deformation (–×–) vs. temperature for cholesterylvinylfumarate polymer/monomer systems with 0.1 p/p polymer. The symbols of the curves relate to data on which the state diagrams of Figures 7 and 8 are based. The shadowed areas were converted to kcal/mol and are represented in Figure 6.

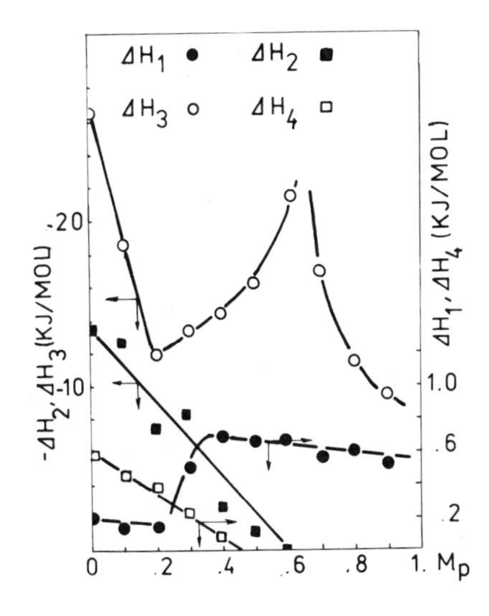

Figure 6. DSC transition heats of cholesterylvinylfumarate polymer/monomer systems

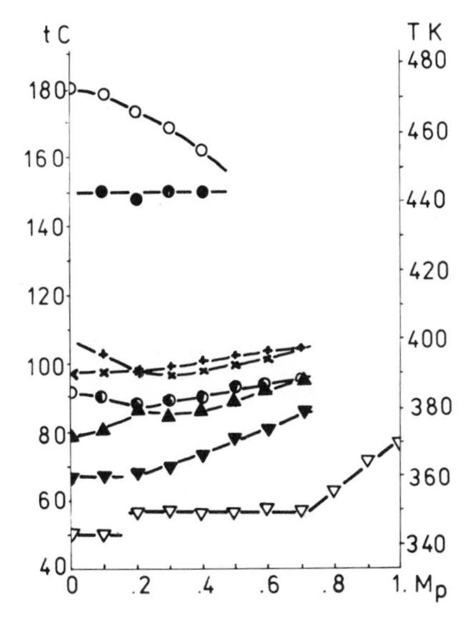

Figure 7. Polymer/monomer state diagram of cholesterylvinylfumarate obtained by DSC. (▽), ΔH_1 initial stage of transition, (▼), second peak; (▲, ◐, ×, and +), peaks or shoulders on the main transition of ΔH_2. (●), ΔH_4; initial stage of transition, (○), second peak on ΔH_4 transition.

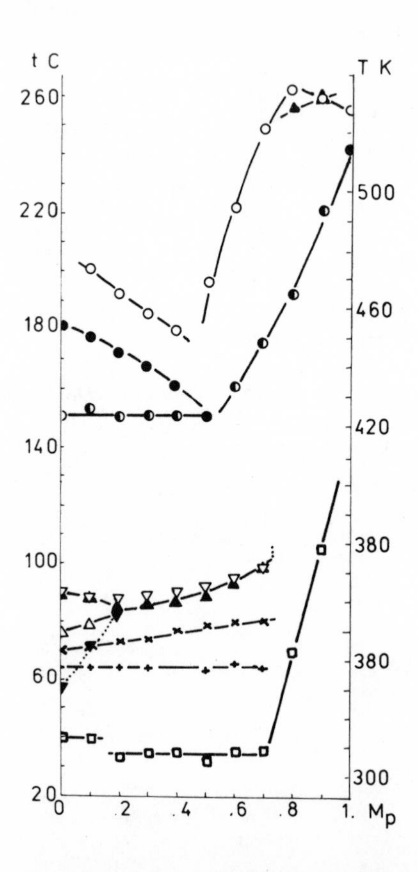

Figure 8. Polymer/monomer state diagram of cholesterylvinylfumarate obtained by thermomechanics and polarizing microscopy. (□), T_g; (+), *initial stage of plastic deformation;* (×), *change in the thermal coefficient of deformation;* (▲), T_f. (▼), *decrease in the birefringence;* (▽), *cholesteric transition, with a gap in the birefringence;* (◐), *decrease in the birefringence;* (●), *peak or shoulder in the depolarized light intensity;* (○), *clearing point.*

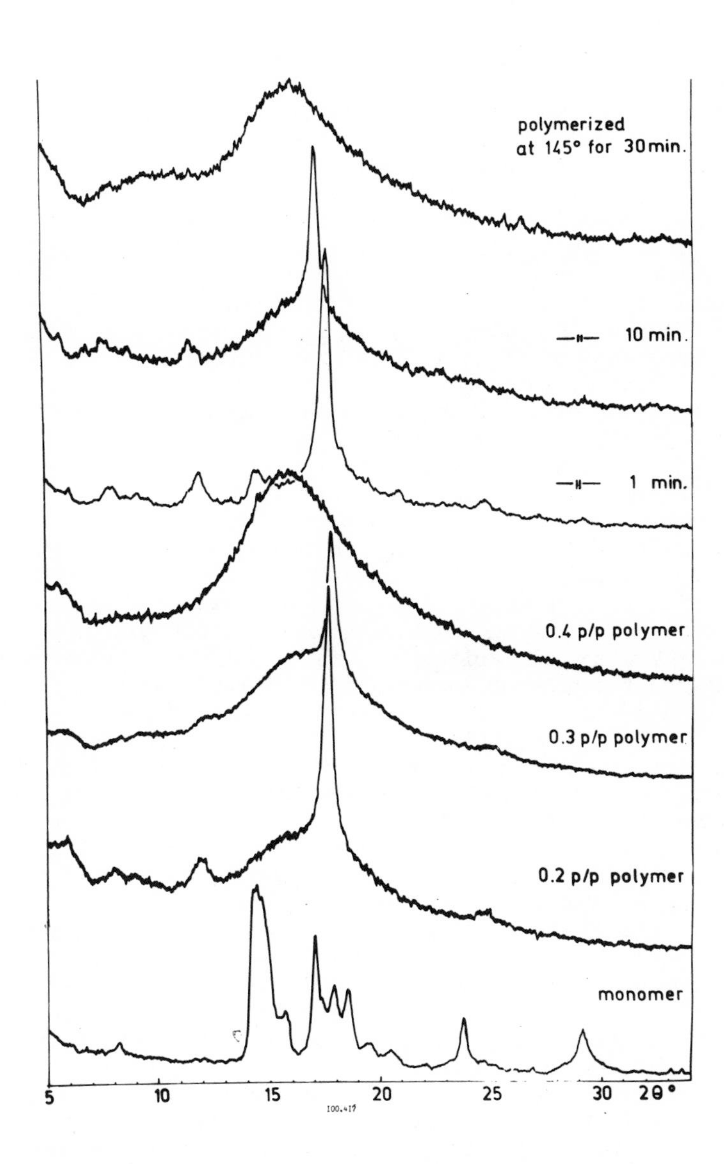

Figure 9. Cu-Kα x-ray diffractograms of cholesterylvinylfumarate polymer/monomer systems prepared by different ways

polymer. When the polymer content is below 70%, the systems contain two phases. Based on thermomechanical observations, the enthalpy change, ΔH_1, is consistent with the T_g of the whole system. The enthalpy change ΔH_2 is associated with the monomer phases containing the polymer, and it consists of several transitions located above the melting temperature, T_f.

The greatest changes in enthalpy are observed at T_f, but they are lower than would have been expected (on the basis of molecular dimensions) from crystalline-to-mesomorphic transitions. The cholesteric state was observed at the flow temperature under a polarizing microscope. The cholesteric reflections are observed at lower temperatures than T_f in systems containing polymers or at T_f of the monomer itself. This means that there is a phase transformation in the cholesteric state too, which corresponds to the last peak of ΔH_2. The dependence of the transitions on concentration is similar to that of the solid solution.

The cholesteric-to-isotropic liquid transition is of an eutectic type. The enthalpy changes, ΔH_4, are related to the cholesteric phases. The composition of the system with the lowest melting point does not coincide with the composition of the homogeneous polymer (48% and 72%, respectively). The transition from cholesteric to isotropic state takes place simultaneously with an exothermic transition (ΔH_3) associated with polymerization. This exothermic transition depends on the polymer/monomer ratio of the samples with a minimum and a maximum value at polymer contents of 20% and 60%, respectively. Since this transition can be found also in the precipitated polymer, it may result from the transition of cyclopolymeric chains characteristic of vinylfumarates (27).

The reflected color of the cooled samples proves that the polymer dissolved in the monomer phase fixes the cholesteric structure of the monomer, which is possible only for a cholesteric homogeneous mixture of the monomer with the polymer. From this solution a smectic B phase monomer is precipitated by cooling, and thus a two-phase system is obtained consisting of a smectic monomer and a cholesteric polymer plasticized by its monomer. At room temperature, therefore, we have a three-phase system: a plasticized polymer with a polymer content of 70%, a polymer phase with a cholesteric structure enriched in monomer, and a monomer phase with a smectic B structure. Consequently, the polymerization which started in the cholesteric state is a homogeneous reaction with a thermodynamically stable homogeneous phase which is solidified by cooling before a phase separation begins. The reaction rate is decreased to a great extent in the highly elastic state of the plasticized polymer, and a limiting conversion is attained.

Figure 10 shows the polymer/monomer state diagram of a copolymer prepared from *p*-methyl, *p*'-acryloyloxyazoxybenzene and cholesterylvinylsuccinate and its original monomer mixture. The

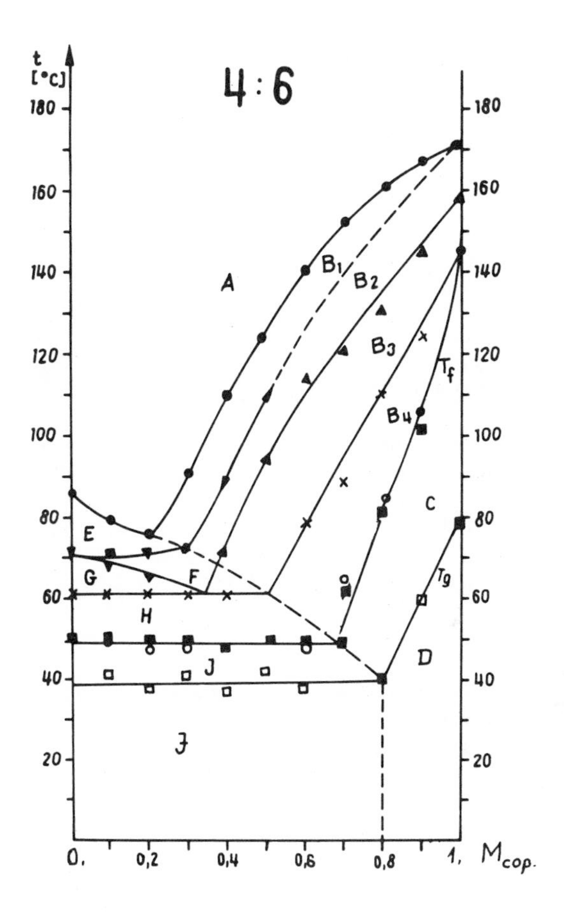

Figure 10. Polymer/monomer state diagrams of a copolymer prepared from a mixture of 0.4 mol of cholesterylvinylsuccinate and 0.6 mol of p-methyl,p'acryloyloxyazoxybenzene and of the original mixture of monomers. (□), T_g; (○), T_f; both obtained by thermomechanics. (■), increase; (×), shoulder; (▼), peak; (▲), constant value of depolarized light intensity; (●), clearing point. The meaning of the areas: (A), isotropic liquid; (B), different smectic phases of the plasticized copolymer in the liquid, (C) in the high elastic, and (D) in the glassy states, respectively. Other areas correspond to biphasic systems, where the plasticized copolymers are in the glassy (J), in highly elastic (I), and in liquid states. (E–H). In addition the mixture of monomers in crystalline (I, J), molten eutectic + crystalline (H), and in the cholesteric (E, F, G) states is given. B_1 and B_3 are biphasic; where one phase is B_2, the other is either isotropic-liquid or B_4.

conditions of mesophase formation are much more complicated here than in the case of homopolymers and will not be discussed in detail. However, the main conclusion that can be made from this and from the other state diagrams of the same monomers with different compositions (28) is that there are at least one or more ranges of temperature and composition where two phases exist with different compositions and different mesophases. Homogeneous phases can be observed at higher polymer contents where the structure is determined by the behavior of the polymer itself and does not depend on the phase of the pure monomer.

Conclusion

In our present work the mesomorphic state polymerization was investigated from the point of view of solid-state polymerization. The thermodynamic interactions of monomers and polymers can be represented by state diagrams. Effects reported by Sadron et al. may only be expected in the homogeneous areas of the state diagrams. The thermodynamic conditions leading to such areas are fulfilled frequently in smectic states but rarely in cholesteric or nematic ones.

Acknowledgement

The authors are indebted to J. Meisel (Technical University of Budapest) for the DSC measurements.

Literature Cited

1. Sadron, C., Pure Appl. Chem. (1962) 4, 347.
2. Bouligand, Y., Cladis, P.E., Liebert, L., Strzelecki, L., Mol. Cryst. Liq. Cryst. (1974) 25, 233.
3. Kaiser, J., Wegner, G., Fischer, E.W., Isr. J. Chem. (1972) 10, 157.
4. Coulson, J.P., Renecker, D.H., J. Appl. Phys. (1970) 41, 4296.
5. Hardy, G., Cser, F., Takács, E., J. Polym. Sci. (1973) C42, 662.
6. Hardy, G., "Kinetics and Mechanism of Polyreactions," Main and Plenary Lectures, Akadémiai Kiadó, Budapest, p. 571, 1970.
7. Hardy, G., Cser, F., Kovács, G., Fedorova, N., Samay, G., Acta Chim. Acad. Sci. Hung. (1973) 79, 143.
8. Wegner, G., Shermann, W., Colloid Polym. Sci. (1974) 252, 655.
9. Hardy, G., Kovács, G., Koszterszitz, G., Fedorova, N., Cser, F., Proc. Tihany Symp. Radiat. Chem. 3rd. 1971 (1972) 601.
10. Cser, F., Hardy, G., Acta Chim. Acad. Sci. Hung. (1975) 84, 297.

11. Kitaigorodskii, A.I., Organic Chemical Crystallography, Consultant Bureau, N.Y., 1961.
12. Hardy, G., Cser, F., Nyitrai, K., Fedorova, N., Proc. Tihany Symp. Radiat. Chem. 4th. (1977) in press.
13. Baughman, R.H., Yee, K.C., J. Polym. Sci. Chem. (1974) 12, 2467.
14. Rosilio, C., Ruaddel-Texier, A., J. Polym. Sci. Chem. Ed. (1975) 13, 2459.
15. Putermann, M., Fort, T., Lando, J.B., Colloid Interface Sci. (1974) 47, 705.
16. Hardy, G., Nyitrai, K., Cser, F., Cselik, G., Nagy, I., Eur. Polym. J. (1969) 5, 133.
17. Hardy, G., Cser, F., Kalló, A., Nyitrai, K., Bodor, G., Lengyel, M., Acta Chim. Acad. Sci. Hung. (1970) 65, 287.
18. Amerik, Y.B., Krentsel, B.A., Dokl. Akad. Nauk. S.S.S.R. (1965) 165, 1097.
19. Hardy, G., Cser, F., Fedorova, N., Bátky, M., Acta Chim. Acad. Sci. Hung. (1977) 94, 275.
20. Paleos, C.M., Labes, M.M., Liq. Crysts. Int. Liq. Cryst. Conf. Pap. 3rd. (1972) 1065.
21. Perplies, E., Ringsdorf, H., Wensdorf, J.H., Makromol. Chem. (1974) 175, 553.
22. Nyitrai, K., Cser, F., Lengyel, M., Seyfried, É., Hardy, G., Eur. Polym. J. (1977) 13, 673.
23. Cser, F., Nyitrai, K., Seyfried, E., Hardy, G., Eur. Polym. J. (1977) 13, 679.
24. Nyitrai, K., Cser, F., Bui Duc Ngoc, Hardy, G., Magy. Kém. Foly. (1976) 82, 210.
25. Nyitrai, K., Cser, F., Csermely, G., Bui Duc Ngoc, Hardy, G., Eur. Polym. J., accepted for publication.
26. de Vries, A., Mol. Cryst. Liq. Cryst. (1973) 24, 337.
27. Yamada, M., Takase, I., Yuki Gosei Kagaku (1963) 20, 180.
28. Cser, F., Nyitrai, K., Kocsis, J., Hardy, G., unpublished data.

RECEIVED April 11, 1978.

9

Applications of Thermotropic Mesophase Reactions

RAJ N. GOUNDER

Corporate Research and Development, Lord Corporation, Erie, PA 16512

Recent research in the areas of molecular/microstructural engineering of polymer solids has resulted in a number of technological breakthroughs. As a result, we are now close to realizing the theoretically possible maximum performance from these microstructurally engineered polymeric materials. Table 1 lists a number of tools that are now available for the microstructural designing of ultimate polymer solids. The potentials and limitations of these various tools for molecular/microstructural engineering of polymer solids are the subjects of a monograph[1] currently under preparation. Some of these technologies offer great potentials for producing polymeric solids with ultimate behaviors in all three crystallographic directions. Thus, the technique of solid state polymerization has shown the practicality of producing defect-free single crystals of chain extended polymeric systems, provided certain stringent criteria may be satisfied[2-5]. Ultimate properties along specific directions only, however, are obtained more easily by a variety of techniques. Thus, a number of techniques utilizing certain extrinsic variables have been very widely investigated. These include the control of mechanical and thermal parameters on polymer solids[6-7], control of mechanical (hydrostatic) and thermal parameters on polymer melts[8-10], and solutions[11-12], as well as techniques utilizing the effects of electric and magnetic fields[13] in polymeric systems. A number of technologies have also been developed utilizing the effects of intrinsic parameters for controlling the microstructural architectures of polymeric solids. Examples of these techniques utilize polymer-solvent interactions[14-15], polymer-polymer interactions[16] and interchain features[17]. Two techniques with maximum success utilize thermotropic mesophase systems and lyotropic liquid crystalline

0-8412-0419-5/78/47-074-108$05.00/0

systems respectively. The Kevlar® fibers were developed from lyotropic liquid crystalline systems[14]; the equally interesting ultra-high modulus graphite fibers are made from thermotropic mesophase pitches. The application of thermotropic liquid crystalline or the so-called mesophase reactions for the control of molecular/ microstructural architectures in macro-molecular systems will be the subject of this review.

Applications of Carbonaceous Mesophase Pitch

The formation of mesophase structures from the liquid pyrolysate during the heat treatment of potentially graphitizable materials such as aromatic hydrocarbons, coal-tar, and petroleum pitches was first reported by Brooks and Taylor[18-19]. The totally isotropic tar or pitch transforms to an almost completely anisotropic mesophase pitch at temperatures ranging from 300-500°C by the parallel alignment of large planar molecules developed by aromatic polymerization at such temperatures. Hot stage microscopic studies[20-21] have revealed that on heating a fully isotropic tar or pitch under quiescent conditions at a temperature of about 350-450°C, small insoluble liquid spheres begin to appear (Figure 1). These spheres gradually increase in size as heating is continued. These spheres are found to be highly anisotropic in character.

Polarized light microscopy as well as electron diffraction studies show that these spheres consist of layers of oriented sheet-like molecules aligned in nearly the same direction (Figure 2). These sheet-like or planar molecules were developed by aromatic polymerization upon heating at temperatures over 300°C (Figure 3). Based upon their electron diffraction and polarized light absorption studies, Brooks and Taylor proposed a model for the anisotropic spheres as shown in Figure 4. According to this model, the layer at the equatorial plane is perfectly flat whereas the layers away from the equatorial plane shows increasing amounts of curvature.

As heating in the range of 300-500°C is continued, these spheres continue to grow until they come into contact with one another, whereupon they gradually coalesce with each other to produce large masses of aligned layers (Figure 5). Thus domains of aligned molecules much larger than those of the original spheres are formed. These domains come together to form a bulk mesophase wherein the transition from one oriented domain to another sometimes occurs smoothly

®Dupont trade mark for polyaramid fibers

TABLE 1

MOLECULAR/MICROSTRUCTURAL ENGINEERING OF POLYMER SOLIDS

Reactions in Pre-Oriented Media
Solid State Reactions
Mesophase Reactions
Thin Layers
Effect of Intrinsic Parameters
Solvent-Polymer Interactions
Polymer-Polymer Interactions
Interchain Features
Effect of Extrinsic Parameters
Mechanical and Thermal Parameters on Polymer Solids
Mechanical and Thermal Parameters on Polymer Melts
Mechanical Parameters on Polymer Solutions
Electric and Magnetic Fields

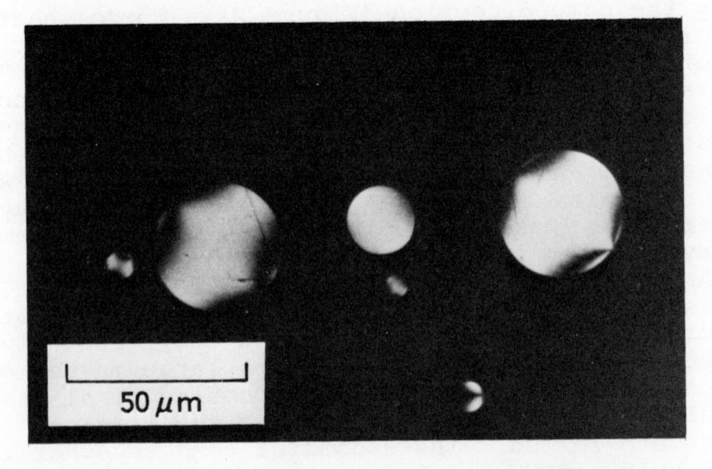

Metallography

Figure 1. Photomicrograph of mesophase spherules (41)

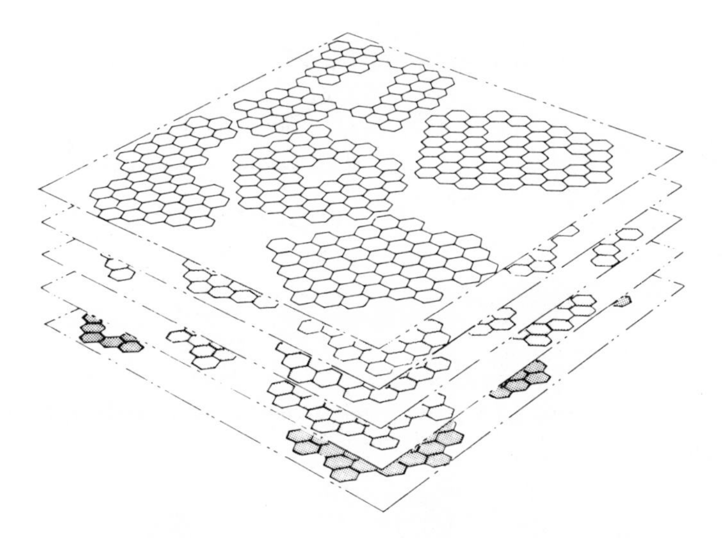

Molecular Crystals and Liquid Crystals

Figure 2. Lamellar liquid crystal model of the carbonaceous mesophase [(1977) 38, 199]

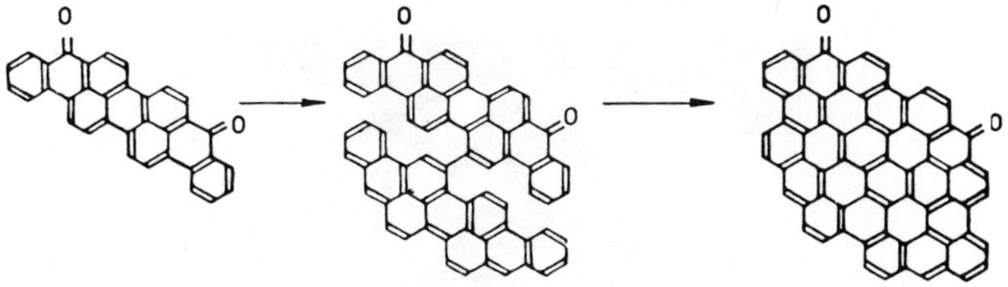

O O O

O O O

(I) (II) (III)

Figure 3. Formation of planar molecules by aromatic polymerization

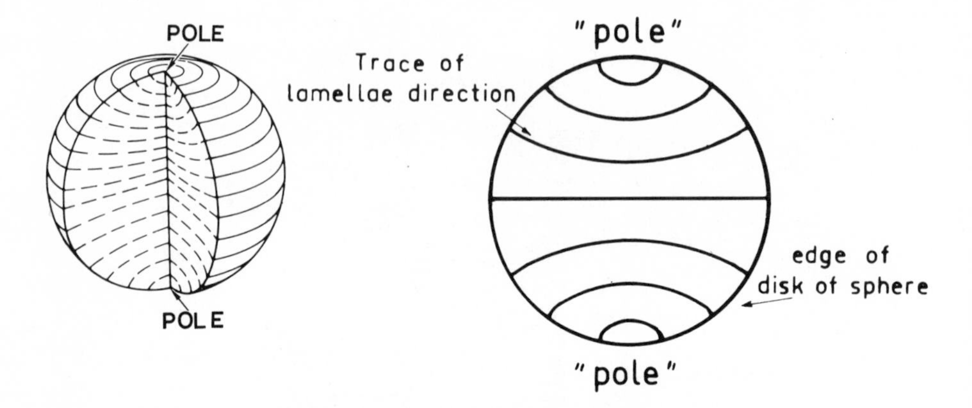

*Figure 4. (*left*) Mesophase sphere with section including polar diameter; (*right*) cross section of mesophase sphere*

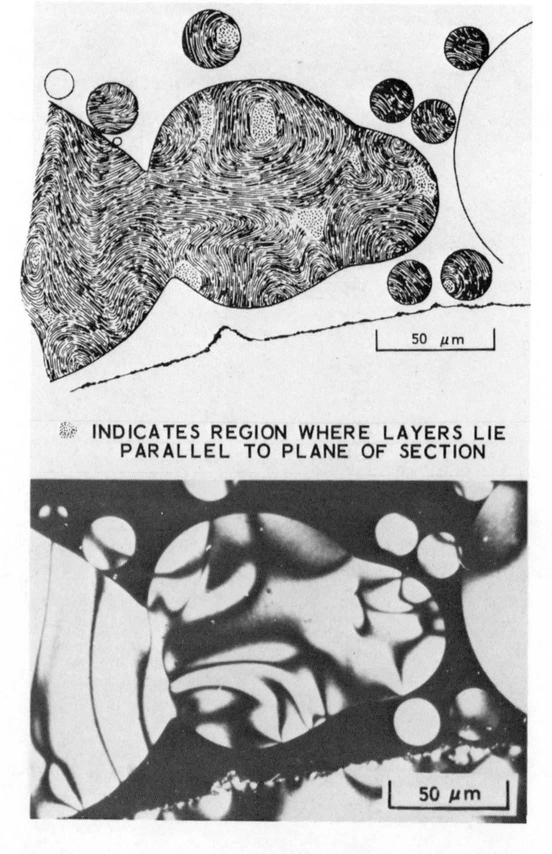

Molecular Crystals and Liquid Crystals

Figure 5. Coalescence of mesophase spheres (1977) 38, 177

and continuously through gradually curving lamellae and sometimes through more sharply curving lamellae. The difference in orientation between the domains creates a complex array of polarized light extinction contours in the bulk mesophase corresponding to various types of linear discontinuity in molecular alignment.

The molecular arrangements within the mesophase pitches thus obtained resemble that of conventional nematic liquid crystals. It has been shown that like conventional nematic liquid crystals[22-25] the mesophase pitches possess plastic properties[26], high degrees of anisotropy[19], magnetic field effects[27], confining surfaces[28], and eutectic effects[29]. However, the mesophase pitches possess other properties which are different from the behaviors of conventional nematic liquid crystals. For example, the mesophase is formed irreversibly; it is relatively insoluble in benzene and pyridine[30]. Mesophase pitches exhibit high activation energies (38-85 K Cal mol^{-1}) which correspond to chemical rather than physical processes[31-32]. Further, increases in molecular weight[19,35] and C/H ratios[21,30] and the kinetics[34] of mesophase formation in pitches all indicate chemical processes of mesophase formation. These are not properties of conventional nematic liquid crystals. Thus, the mesophase pitches while being like nematic liquid crystals in the structures they possess, are unlike nematic phases in their being formed through chemical processes.

Since the discovery of mesophase pitches by Brooks and Taylor in 1965, numerous studies have been carried out to understand and optimize the effects of various parameters on the mesophase nucleation, growth, and coalescence. Table 2 lists a number of parameters that effect the formation of mesophase spheres - their nucleation, growth and coalescence as well as the properties of the resulting mesophase pitch. For example, use of high pressures during heat treatment to produce mesophases has been shown[35] to inhibit mesophase coalescence and produce a liquid crystal system composed of uncoalesced spheres. Mesophase sphere growth and coalescence are also hindered by the presence of insoluble carbon particles[31] and metal oxide particles[36]. Of course, the structure and size of mesophases obtained by the heat treatment of carbonaceous pitches very much depend upon the chemistry of parent substances[37] and the nature of radical intermediates of the pyrolytic process[38]. The amount and the composition of the mesophase as well as any co-existing isotropic phase in a pyrolysate pitch depend upon temperature[39-41]. The viscosity of the pyrolysate very much controls the coalescence of mesophase spheres and hence the formation of anisotropic domains. Various studies[39-42] have been performed to study the effects of temperature, heating time and viscosity on the mesophase sphere formation, growth and coalescence. Yamada, et al[43] have shown that removal of low molecular weight molecules from partially transformed pitches by solvent extraction affects the viscosity and hence prevents sphere coalescence during additional heating. Thus, a body of knowledge has been developed exploring the effects of various parameters on the nature of mesophases formed in carbonaceous pitches.

During this same period, the technology of producing carbon fibers from various pitches was also being established.[44-50] The already existing body of knowledge concerning mesophase formation in pitches has greatly impacted on the development of pitch precursor carbon fiber technology. Thus, about the same time Brooks and Taylor of Australia published their results on the mesophase formation in carbonaceous pitches, Otani[51] of Japan published his work on the melt spinning of carbon fibers from conventional isotropic carbonaceous pitches. Otani's process involved several steps as shown in Figure 6. First, carbon fibers were spun from pitches specially prepared by pyrolyzing polyvinyl chloride at a temperature of 400-415°C for 30 minutes in a nitrogen atmosphere. The melt-spun fiber was made infusible by oxidizing in ozone or air and was then carbonized at a temperature in the range of 500-1350°C. The fibers prepared in this manner exhibited a tensile modulus and strength of 8×10^6 psi and 256×10^3 psi respectively. Later[52], carbon fibers with similar properties were also prepared from naturally occuring pitches such as petroleum asphalt and coal-tar pitches.

A few years later, Hawthorne and coworkers[53] in Canada showed that the mechanical properties of pitch fibers could be greatly enhanced by stretching the fiber during the carbonization step and carrying out the carbonization at a much higher temperature of 2000-2800°C (see Figure 7). The fibers thus obtained by straining during carbonization exhibited an elastic modulus as high as 70×10^6 psi and a tensile strength of almost 375×10^3 psi. Such fibers were shown to exhibit a high degree of crystallite orientation parallel to the longitudinal fiber axis. However, the individual crystallites were found to be turbostratic and essentially devoid of three-dimensional order charac-

TABLE 2

EFFECT OF PARAMETERS ON MESOPHASE FORMATION

1. Pressure
2. Insoluble Particles
3. Temperature
4. Parent Material
5. Nature of Radical Intermediate
6. Viscosity
7. Heating Rate and Time

Isotropic Pitch $\xrightarrow[\text{Spinning}]{\text{Melt}}$ Spun Fiber $\xrightarrow[\text{in Ozone}]{\text{Oxidize}}$ Infusible Fiber $\xrightarrow[\text{500-1350° C}]{\text{Carbonize}}$ Carbon Fiber

$E_T = 8 \times 10^6$ psi

$\sigma_T = 256 \times 10^3$ psi

Carbon

Figure 6. Fibers from isotropic pitches (51)

Isotropic Pitch $\xrightarrow[\text{Spinning}]{\text{Melt}}$ Spun Fiber $\xrightarrow[\text{in Ozone}]{\text{Oxidize}}$ Infusible Fiber $\xrightarrow[\text{2000 - 2800° C}]{\textit{Strain}\ \text{\& Carbonize}}$ Oriented Carbon Fiber

$E_T = 70 \times 10^6$ psi

$\sigma_T = 375 \times 10^3$ psi

Oriented Fibers

Turbostratic

Devoid of 3D Order

Nature

Figure 7. Fibers from strained isotropic pitches (53)

teristic of polycrystalline graphite.

Only recently, the usefulness of melt-spinning carbon fibers from a carbonaceous pitch in its mesophase state has been realized and patented[54-59]. According to these patents, petroleum pitch, coal tar pitch or acenaphthylene pitch is first converted to the mesophase state by heating under quiescent conditions and under vacuum or nitrogen atmosphere at a temperature of about 350-450°C (see Figure 8). The carbonaceous pitch having a mesophase content of 55 to 65 percent thus produced is spun into fibers by conventional techniques such as melt spinning, centrifugal spinning or blow spinning. For 55 to 65 percent mesophase content, suitable spinning temperatures range from 340°C to 380°C. The thermoplastic carbonaceous fibers thus obtained are given a thermosetting treatment by heating in an oxygen-containing atmosphere at temperatures below the softening point of the fibers but above at least 250°C for 5 to 60 minutes. The infusible fibers obtained by thermosetting are then carbonized by heating in an inert atmosphere at a temperature ranging from 1500°C to 1700°C for about 1 to 5 minutes. The carbonized fibers are further heat treated in an inert atmosphere at a temperature of 2800°C to 3000°C for about 1 minute. The fibers thus obtained possess the three dimensional order of polycrystalline graphite. The fibers consist of highly oriented graphitic domains of 5000 A to 40,000 A in size. As a result, the fibers obtained exhibit extremely high mechanical properties with the Young's modulus in the range of 75×10^6 psi to 120×10^6 psi and tensile strength in the range of 250×10^3 psi, to 350×10^3 psi.

Besides the applications in the production of ultrahigh modulus carbon fibers, the mesophase pitches are used in making thermal-shock-resistant aerospace graphites[60] and graphite electrodes[61] for the steel industry. Needle cokes are produced by deformation through turbulence and bubble percolation of carbonaceous mesophase pitches[60,62]. Such anisotropic needle cokes form filler materials in the production of thermal-shock-resistant aerospace graphites and graphite electrodes. The needle cokes are shown to consist of acicular-shaped, corrugated aromatic mesophase layers that are tightly folded and entwined. Increased content of the acicular phase in graphites is shown to decrease the thermal expansion[63-65] and increase the resistance to cleavage and fracture[60].

Development of Liquid Crystalline Polyesters

Another recent development involving the use of thermotropic mesophases is the discovery of polyesters exhibiting liquid crystalline melts. These polyesters are prepared by synthesizing polymers containing moieties known to lead to liquid crystallinity in non-polymeric materials. The structural requirements for the liquid crystal formation has been well-studied by Gray[22]. Most conventional nematic and smectic liquid crystals approximate a general structure as shown in Figure 9. Figure 9 also shows a few examples of the linkage groups and terminal groups that facilitate the formation of liquid crystals. The recent development of liquid crystalline polyesters make use of such moieties that are known to facilitate liquid crystal formation.

Thus, Kuhfuss and coworkers[66-69] successfully prepared a list of polyesters containing moieties listed in Figure 10. Many of these polyesters and copolyesters are found to be liquid crystalline in their molten state. For example, these workers prepared a copolyester of PET with a number of liquid crystal forming moieties. These copolyesters, in addition to the PET segments III, contained the liquid crystal favoring segments II and IV (see Figure 11). These copolyesters were prepared by the acidolysis of PET with p-acetoxybenzoic acid and polycondensation through the acetate and carboxyl groups. Thus a list of copolyesters containing various mole-fractions of para oxybenzoyl moieties were obtained.

It was found that melts of copolyesters containing 40-90 mole percent para oxybenzoyl segments exhibited behaviors similar to those of nematic liquid crystals. As shown in Figure 12, measurements of the melt viscosities of copolyesters containing varying mole percent para oxybenzoyl revealed very interesting features. Up to 30 mole percent para oxybenzoyl content the melt viscosity increased. However, higher than 30 mole percent para oxybenzoyl resulted in a decrease in the melt viscosity leading to a minimum around 60 mole percent para oxybenzoyl. Further, copolyesters containing in the range of 0-30 mole percent para oxybenzoyl exhibited clear melts, whereas copolyesters containing greater than 40 mole percent para oxybenzoyl led to opaque melts. This was a clear indication of liquid crystal formation at greater than 40 mole percent para oxybenzoyl contents.

The copolyesters processed from such liquid crystal forming melts exhibit very interesting mechanical properties. Figure 13 shows the effect of para oxybenzoyl content on the mechanical properties (in the

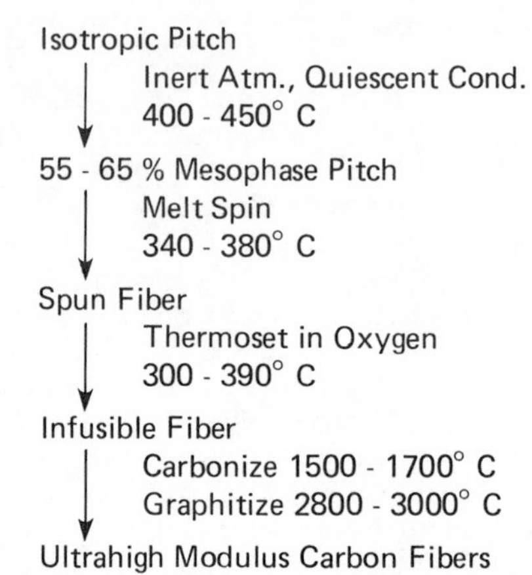

U.S. Patent 4,005,183

Figure 8. Carbon fibers from mesophase pitches (58)

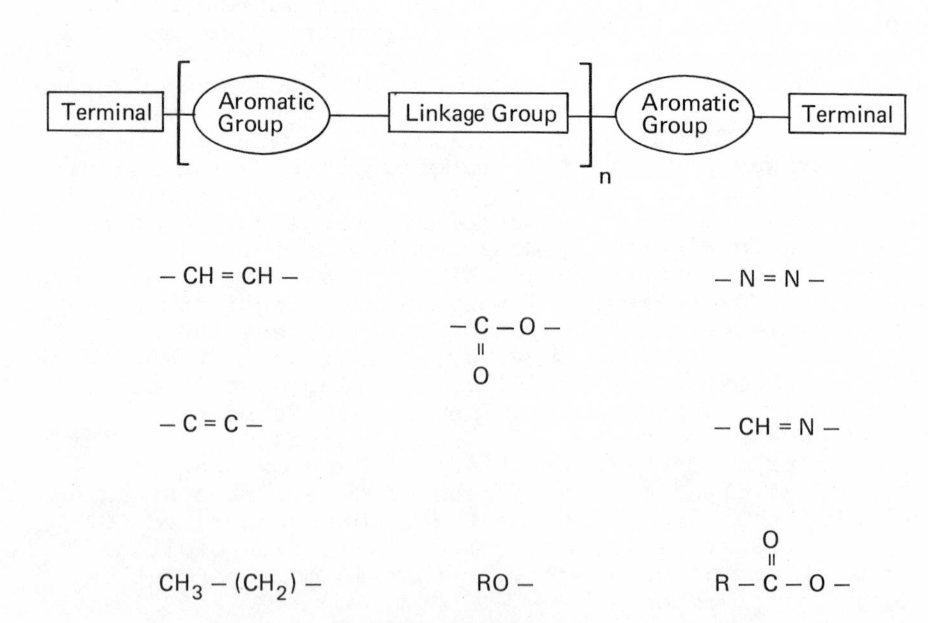

Figure 9. (top) Structure of liquid crystals; (middle) examples of linkage groups; (bottom) examples of terminal groups

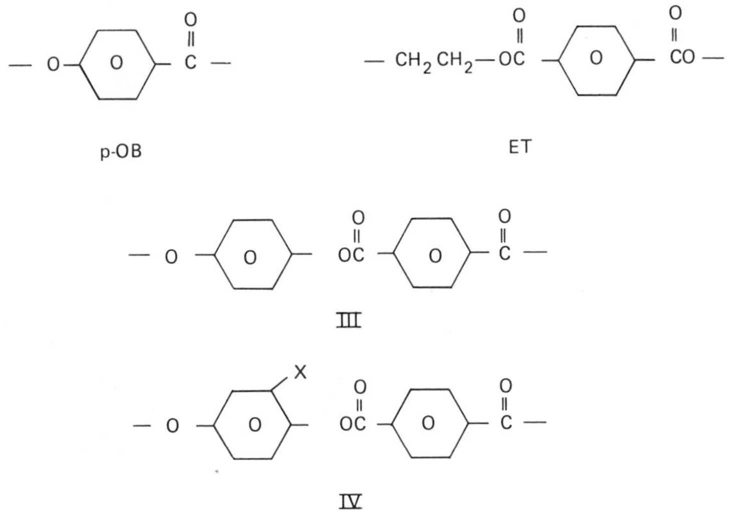

Figure 10. Polymers containing moieties known to lead to liquid crystallinity

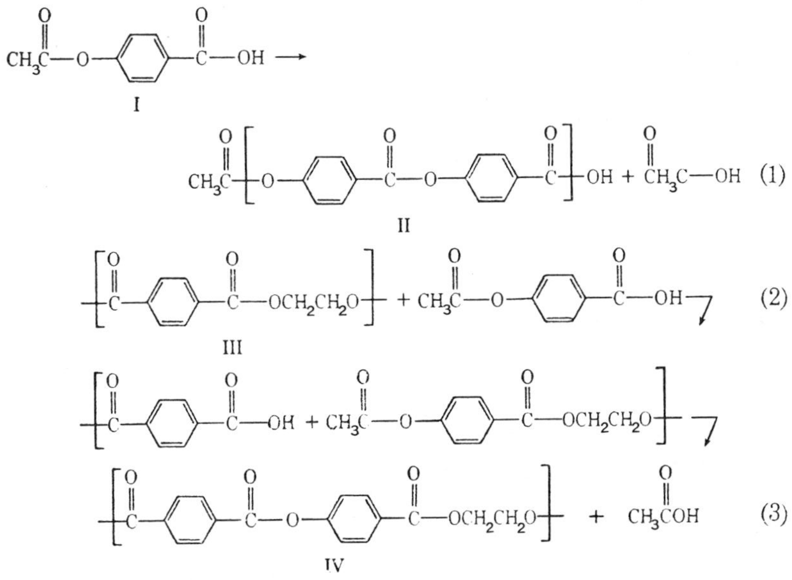

Figure 11. Synthesis of copolyesters containing liquid crystalline segments

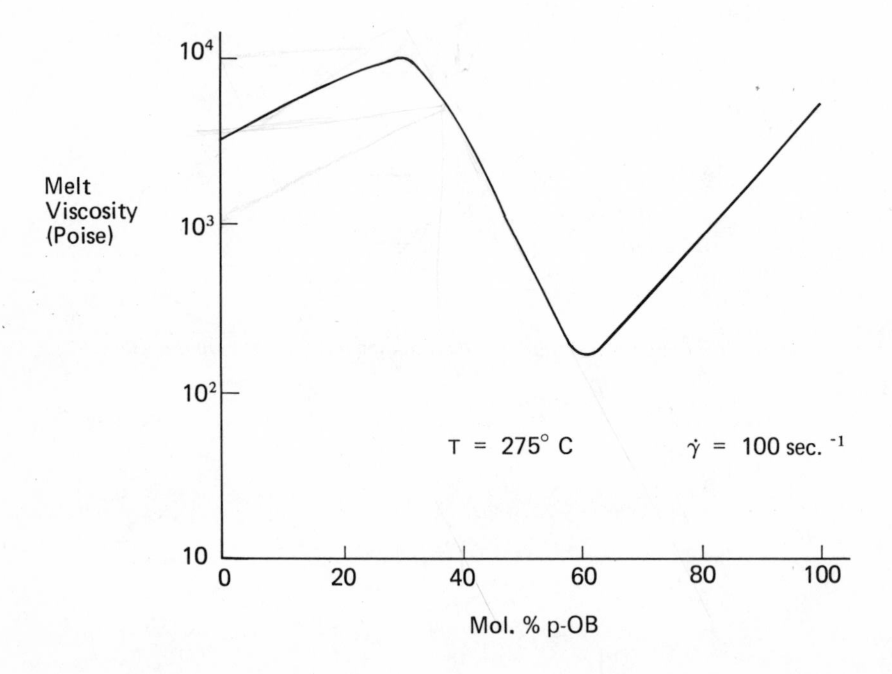

Figure 12. Effect of p-*OB content on the melt viscosity of copolyesters*

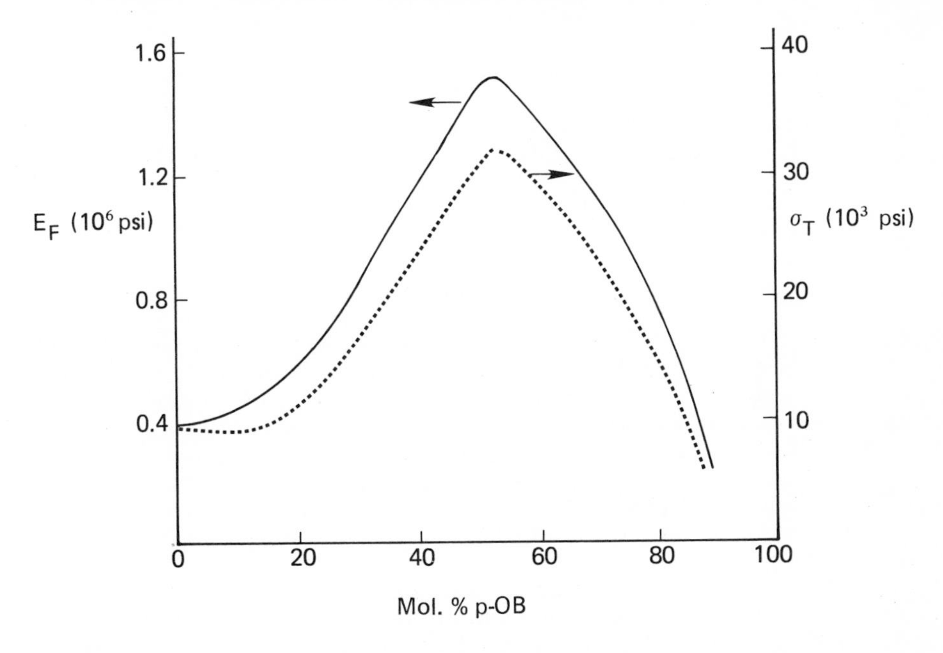

Figure 13. Effect of p-*OB content on the mechanical properties of copolyesters*

TABLE 3

MECHANICAL PROPERTIES OF LIQUID CRYSTALLINE POLYMER CONTAINING 60 MOL. % p-OB

	Transverse	Longitudinal
E_F (10^6 psi)	0.23	1.7
σ_F (10^3 psi)	4.9	15.9
σ_T (10^3 psi)	4.2	15.5
Break Elongation (%)	10	8
Notched Izod Impact (ft.-lb./in.)	0.6	6.1

injection molding direction) of injection molded copolyesters. The continuous line represents the flexural modulus and the dotted line denotes the tensile strength. The plot indicates that maximum in these properties are achieved around 60 mole percent para oxybenzoyl content, thus indicating the effect of liquid crystal formation on the mechanical properties of resulting polymer solid.

Table 3 summarizes the mechanical properties of an injection molded liquid crystalline polymer containing 60 mole percent para oxybenzoyl. The mechanical properties were measured parallel to the injection molding direction (longitudinal) as well as transverse to the injection molding direction (transverse). The flexural moduli, tensile strengths, elongation to break and impact strength of copolyesters exhibiting liquid crystalline melts are found to be similar to or higher than those of commercial glass fiber reinforced polyesters.

McFarlane and coworkers[69] have synthesized other polymers containing other moities that are also known to lead to liquid crystallinity in non-polymeric forms. Thus, a wide variety of high molecular weight copolymers exhibiting liquid crystalline melts may be prepared. Structural shapes consisting of a high degree of molecular orientations and chain extensions may be easily fabricated from such liquid crystalline melts.

Polymers with Liquid Crystalline Side Chains

Extensive work has been done in a number of laboratories[70-75] on the synthesis and characterization of several polymers with liquid crystalline side chains. Various of the papers in this monograph deal with various aspects of such polymers containing mesomorphic side chains. A review of the potential applications of such side chain liquid crystalline polymers therefore, will be redundant.

Conclusion

In conclusion, it may be said that the thermotropic liquid crystalline reactions offer a highly efficient technique for the control of the molecular/microstructural architectures in macromolecular systems. A variety of novel polymeric systems exhibiting unique mechanical and physical properties have been possible through special molecular/microstructural designs obtained by utilization of thermotropic mesophase reactions.

Literature Cited

1. R. N. Gounder and S. H. Carr, "Molecular/Microstructural Engineering of Polymer Solids", in preparation.
2. R. H. Baughman and K. C. Yee, J. Macromol. Sci. - Review, in press.
3. R. H. Baughman, "Proceedings of the 8th Biennial Polymer Symposium of Division of Polymer Chemistry", ACS (Plenum Press, 1977).
4. D. Bloor, L. Koski and G. C. Stevens, J. Materials Sci., 10, 1689 (1975).
5. G. Wegner, E. W. Fischer and A. Munoz - Escalona, Die Makromol. Chem., Supl. 1, 521 (1975).
6. E. S. Clark and L. S. Scott, Polymer Engrg. & Sci., 14, 682 (1974).
7. T. Williams, J. Materials Sci., 8, 59 (1973).
8. R. S. Porter and J. H. Southern, J. Makromol. Sci. - Phys., B4, 541 (1970).
9. D. C. Bassett and G. E. Atenburrow, J. Materials Sci. , 12, 192 (1977).
10. A. Keller and P. J. Barham, J. Materials Sci., 11, 27 (1976).
11. A. J. Pennings and P. Smith, J. Materials Sci., 11, 1450 (1976).
12. A. J. McHugh and E. H. Forrest, J. Polymer Sci.: Polymer Physics Ed., 13, 1643 (1975).
13. M. S. Akutin et al, Plasticheskie Massy, 11, 73 (1975).
14. S. L. Kwolek, U. S. Patent 3,671,542 assigned to DuPont (June 20, 1972).
15. B. K. Daniels, J. Preston and D. A Zaukelis, U.S. Patent 3,600,269, assigned to Monsanto (August 17, 1971).
16. T. I. Ablazova et al, J. Appl. Polymer Sci., 19, 1781 (1975).
17. F. E. Arnold and R. L. Van Duesen, J. Appl. Polymer Sci., 15,2035 (1971).
18. J. D. Brooks and G. H. Taylor, Nature, 206, 697 (1965).
19. J. D. Brooks and G. H. Taylor, Carbon, 3, 185 (1965).
20. R. T. Lewis, "Extended Abstracts of 12th Biennial Conference on Carbon", (The American Carbon Society, 1975), pp. 215.
21. D. O. Rester, ibid, pp. 227.
22. G. W. Gray, "Molecular Structure and the Properties of Liquid Crystals" (Academic Press, London &

New York, 1962).

23. E. B. Priestley, P. J. Wojtowicz and P. Sheng, "Introduction to Liquid Crystals", (Plenum Press, New York & London, 1975).
24. D. G. deGenner, "The Physics of Liquid Crystals", (O.U.P., 1974).
25. G. H. Brown et al, "A Review of Structure and Properties of Liquid Crystals", (Butterworths, 1971).
26. J. L. White, Aerospace Report TR-0074 (4250-40)-1, (1974).
27. Y. Sanada et al, Carbon, 10, 644 (1972).
28. H. Marsh, Fuel, London, 52, 205 (1973).
29. H. Marsh, et al, Fuel, London, 53, 168 (1974).
30. K. J. Huttinger, Bitumen, Teere, Asphalte, 24, 255 (1973).
31. H. Honda et al, Carbon, 8, 181 (1970).
32. I. C. Lewis and G. H. Jackson, "Extended Abstracts of 12th Biennial Conference on Carbon" (The American Carbon Society, 1975), pp. 267.
33. J. D. Brooks and G. H. Taylor, "Chemistry and Physics of Carbon" edited by P. L. Walter, Vol. 4, pp. 243-286 (Marcel Dekker, New York, 1968).
34. I. C. Lewis and L. S. Singer, "Extended Abstracts of the 12th Biennial Conference on Carbon" (The American Carbon Society, 1975), pp. 265.
35. H. Marsh et al, Carbon, 9, 159 (1971).
36. E. Fitzer et al, "International Carbon Conference", (Baden-Baden, West Germany, 1972), Paper Ch. 5.
37. H. Marsh et al, "Extended Abstracts of 12th Biennial Conference on Carbon" (The American Carbon Society, 1975), pp. 117.
38. L. G. Isaacs, Carbon, 8, 1 (1970).
39. M. Ihnatowicz et al; Carbon 4, 41 (1966).
40. D. O. Rester and C. R. Rowe, Carbon, 12, 218 (1974).
41. J. DuBoise, C. Agace and J. L. White, Metallography, 3, 337 (1970).
42. J. L. White and R. J. Price, Carbon, 12, 321 (1974).
43. Y. Yamada et al, Carbon, 12, 307 (1974).
44. S. Otani, U. S. Patent 3,392,216 assigned to Kureha Kagaku Kogyo Kabushiki Kaisha (July 9, 1968).
45. T. Ishikawa et al, U.S. Patent 3,552,922 assigned to Nippon Carbon Co., Ltd., (January 5, 1971).
46. L. A. Joo et al, U. S. Patent 3,595,946 assigned to Great Lakes Carbon Corp.,(July 27, 1971).
47. S. Otani, U. S. Patent 3,629,379 assigned to Kureha Kogyo Kabushiki Kaisha (December 21, 1971).
48. T. Araki et al, U. S. Patent 3,702,054 assigned to Kureha Kagaku Kogyo Kabushiki Kaisha (November 7, 1972).

49. L. A. Joo et al, U.S. Patent 3,718,493 assigned to Great Lakes Carbon Corp. (Feb. 27, 1973).
50. M. Toyoguchi et al, U. S. Patent 3,767,741 assigned to Mitsubishi Oil Co. (October 23, 1973).
51. S. Otani, Carbon, 3, 31 (1965).
52. S. Otani, et al, Carbon, 4, 425 (1966).
53. H. M. Hawthorne et al, Nature, 227, 946 (1970).
54. L. I. Grindstaff and M. P. Whittaker, U. S. Patent 3,787,541 (January 22, 1974).
55. R. C. Stroup, U. S. Patent 3,814,566 assigned to Union Carbide Corp. (June 4, 1974).
56. D. A. Schulz, U. S. Patent 3,919,376 assigned to Union Carbide Corp. (November 11, 1975).
57. I. C. Lewis, U. S. Patent 3,995,014 assigned to Union Carbide Corporation (November 30, 1976).
58. L. S. Singer, U. S. Patent 4,005,183 assigned to Union Carbide Corp. (January 25, 1977).
59. R. Didchenko et al, "Extended Abstracts of 12th Biennial Conference on Carbon" (The American Carbon Society, 1975), pp. 325.
60. J. E. Zimmer and J. L. White "Extended Abstracts of 12th Biennial Conference on Carbon" (The American Carbon Society, 1975), pp. 223.
61. O. Wegener et al, U. S. Patent 3,761,387 assigned to Rutgerswerke Aktiengesellschaft (Sept. 25, 1973).
62. Y. Suetsugu et al, U.S. Patent 3,799,865 assigned to Nittetsu Chemical Industrial Co. Ltd. (March 26, 1976).
63. S. Mrozowski, "Proc. 1st and 2nd Confs. Carbon" (Waverly Press, 1953), pp. 31.
64. G. Pietzka, G. Wilhelmi and H. Pauls, "International Carbon Conference" (Baden-Baden, West Germany, 1972), pp. 362.
65. A. L. Sutton and V. C. Howard, J. Nuclear Mater. 7, 58 (1962).
66. H. F. Kuhfuss and W. J. Jackson, Jr. U. S. Patent 3,778,410 assigned to Eastman Kodak Co. (Dec. 11, 1973).
67. H. F. Kuhfuss and W. J. Jackson, Jr., U. S. Patent 3,804,805 assigned to Eastman Kodak Co. (April 16, 1974).
68. W. J. Jackson, Jr., and H. F. Kuhfuss, J. Polymer Sci. Polymer Chem. Ed., 14, 2043 (1976).
69. F. E. McFarlane, V. A. Nicely, and T. G. Davis, "Proceedings of the 8th Biennial Polymer Symposium of Division of Polymer Chemistry", ACS (Plenum Press, 1977).
70. A. Blumstein, "Proceedings of the 3rd Midland Macromolecular Conference on Polymerization in

Oriented Systems" (Gordon and Breach, New York, 1977).
71. L. Strzelecki, L. Liebert and P. Keller, Bull. Soc. Chem. 11-12, 2750 (1975).
72. Y. B. Amerik et al, Khim. Volokna, 5, 67 (1975).
73. H. Ringsdorf et al, Die Makromol. Chemie, 176, 2029 (1975).
74. A. C. deVisser et al, Die Makromol. Chemie, 176, 495 (1975).
75. Y. Tanaka et al, Polymer Letters, 10, 261 (1972).

RECEIVED December 8, 1977.

10

Nematic Polymers: Excluded Volume Effects

Y. KIM and P. PINCUS

Department of Physics, University of California, Los Angeles, CA 90024

I. Introduction

There exists a mushrooming interest in the mesomorphic behavior of macromolecules in solution. The existence of macroscopic anisotropy requires (at least partial) chain rigidity. This may arise from next nearest neighbor steric restrictions for polymers constructed from bulky backbone monomers as in the polyamides,[1] e.g.,

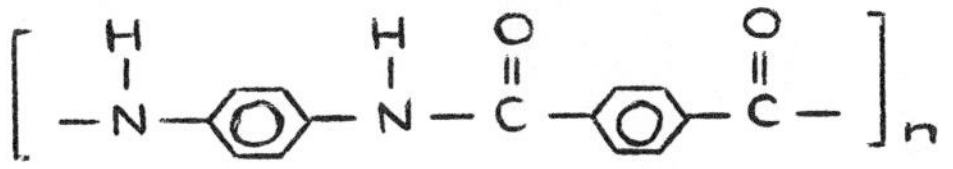

The current symposium, to a great extent, has focused on comb-like systems[2,3] wherein mesogenic side groups are attached to a flexible backbone, as

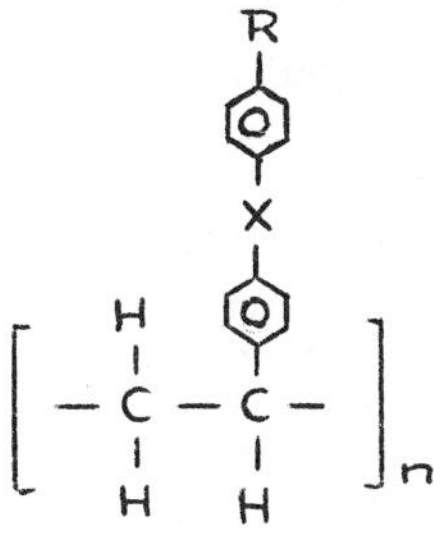

where x is a rigid linkage such as -C = N- and the end group R may be of the form $-O-C_mH_{2m+1}$. The present work mainly concerns helix-coil systems as exemplified by polypeptides, e.g., Poly-Benzyl-L-Glutamate (PBLG)[4,5]

Supported in part by the National Science Foundation and the Office of Naval Research.

0-8412-0419-5/78/47-074-127$05.00/0

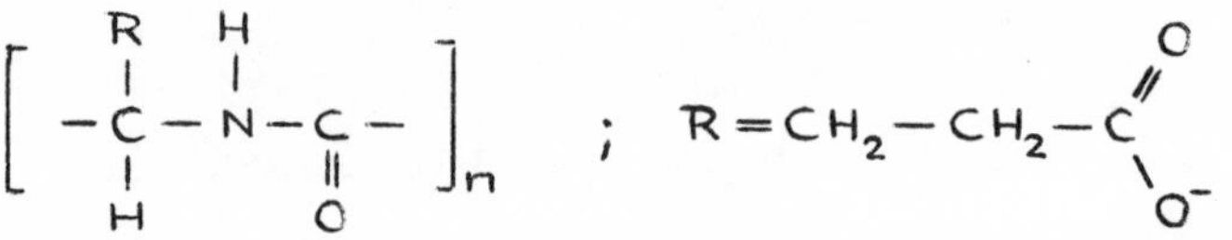

$$\left[-\overset{\overset{\displaystyle R}{|}}{\underset{\underset{\displaystyle H}{|}}{C}} - \overset{\overset{\displaystyle H}{|}}{N} - \underset{\underset{\displaystyle O}{\|}}{C} - \right]_n \; ; \; R = CH_2 - CH_2 - C(=O)O^-$$

in solvents such as dichloroacetic acid or dioxane. Under certain conditions of temperature, pH, ionic strength, etc., hydrogen bonding occurs along the polypeptide backbone between monomers separated by 4 or 5 units leading to a rigid helix. Under such conditions, we are essentially dealing with a solution of rigid rods. If the solvent is good, the only important intermolecular interactions are then steric short range excluded volume repulsions. The mesomorphic behavior of a solution of rigid rods has been studied by several authors.[6,7,8] In particular, for long rods of length b and radius a, Onsager[6] has shown that the effective excluded volume per rod in the isotropic phase is of order b^2a. The actual volume occupied per rod is $\pi a^2 b$. A lyotropic transition to a nematic phase occurs at a rod concentration σ^* given by

$$\sigma^* b^2 a = k \qquad (I.1)$$

where k is a constant in the range 3-10. This implies that for volume fractions in excess of $\pi k(a/b)$ the system should be in an ordered phase. Typical axial ratios[4] for PBLG are $b/a \sim 10^2$ leading to critical volume fractions in the range 10-40%. Indeed such phase transitions are often observed. Our interest here, however, is to consider a somewhat more delicate situation where the isolated polypeptide chains are in essentially random coil conformations. We then suggest that as the concentration of chains increases, a phase transition may occur wherein the molecules become _simultaneously_ rigid and mesomorphic. This _induced rigidity_ may arise when the reduction of the excluded volume energy that obtains in the nematic state of rigid rods balances the decrease in entropy associated with the loss of chain flexibility. This may occur for concentrations in excess of that associated with σ^* of prestretched chains. The goal of the present study is to investigate the phase boundary separating ordered and isotropic phases, as a function of concentration and parameters relating to chain rigidity. In particular, this preliminary report will be devoted to the limit where the isolated macromolecules have extremely sharp helix-coil transitions.

In Section II, we shall briefly recall the statistical mechanics of the helix-coil transition for isolated chains. We shall then present an argument showing that there is little tendency for non-interacting molecules to exhibit substantial mesomorphic fluctuations. A mean field theory for incorporating the interchain interactions will be developed in Section III in the cooperative limit for helix-coil transition. Some final remarks will be made in Section IV.

II. Isolated Molecules

The helix-coil transition of a polypeptide chain consisting of N monomers is usually described in terms of two parameters:

$$s = \exp[-\Delta F_1/k_B T] \tag{II.1}$$

where $-\Delta F_1$ is the gain in free energy per monomer when it is part of a rigid helical segment;

$$\Sigma = \exp[-2\Delta F_2/k_B T] \tag{II.2}$$

where ΔF_2 is the cost in free energy to create a boundary separating helical and randomly coiled segments. For an Ising model with only nearest neighbor interactions $\Delta F_2 = 0$ ($\Sigma = 1$); on the other hand, for polypeptides with finite range hydrogen bonding interactions, Σ may typically be rather small in the range $10^{-2} - 10^{-4}$. Small values of the cooperativity parameter, Σ, favor sharp helix-coil transitions which necessarily have very few segments. In terms of these parameters, the free energy per macromolecule may be simply written as[9,10]

$$F_o = -Nk_B T \ln \lambda ; \tag{II.3}$$

$$\lambda \cong \frac{1}{2} \{ (1+s) + [(1-s)^2 + 4\Sigma s]^{1/2} \} \tag{II.4}$$

The role of Σ in determining the sharpness of the transition may be graphically seen by considering the persistence length, ν, which is the average number of monomers per helical segment,

$$\nu = \frac{d \ln \lambda/ds}{d \ln \lambda/d\Sigma} - 1 \tag{II.5}$$

In Figure 1, we sketch the persistence length as a function of s for several values of Σ. For small values of the cooperativity parameter it is too costly to have very many domain walls and, thus, the system tends to a situation of a completely random coil or a rigid helix depending on whether s is smaller or larger than unity.

Let us consider an isolated chain with persistence length ν. Such a system may be viewed as a random coil of rods of length νa (a is a characteristic monomer dimension). The question arises of whether or not there is any significant orientational ordering among the segments for the isolated chain. For a polymerization index N, the number of rods is (N/ν) leading to a coil radius (for a Gaussian chain)

$$R \simeq (N/\nu)^{1/2} \nu a \approx N^{1/2} \nu^{1/2} a \tag{II.6}$$

The (N/ν) rigid segments are then confined to a volume of order R^3, giving a rod density

$$\sigma \simeq (N/\nu)\ R^{-3} \simeq (N\nu^5)^{-1/2}\ a^{-3} \tag{II.7}$$

which is small compared to the Onsager critical concentration σ^*,

$$\sigma/\sigma^* \simeq (N\nu)^{-1/2} << 1 \quad . \tag{II.8}$$

Excluded volume interactions will further decrease the density.[11,12] This is a strong argument suggesting that overlapping coils (semi-dilute solutions are necessary to obtain liquid crystalline order in solutions of worm-like macromolecules.

III. Mean Field Theory

We now turn to the development of a crude approximation for the case of semi-dilute solutions of heliogenic macromolecules where the interchain interactions are important. We envisage the molecules at some fixed values (s,Σ) as linear arrays of alternating rigid and coiled segments. The mean number of monomers per helical segment is ν (II.5) and the number of such segments is given by

$$n = (\Sigma/\lambda)(\partial\lambda/\partial\Sigma) \tag{III.1}$$

We assume, for this model, that only excluded volume type interactions are important. We then approximate the rigid rod Onsager interaction[6] by the lowest angular dependent term in its multipole expansion which is of quadrupolar symmetry:

$$U_{ij} = -\varepsilon k_B T \nu_i \nu_j Q_i Q_j \quad , \tag{III.2}$$

ε is a dimensionless constant, $\nu_{i,j}$ are the numbers of monomers on the ith and j^{th} segments; Q_i is a quadrupole component associated with the orientation of each segment

$$Q = \frac{1}{2}\ (3\ \cos^2\theta - 1) \tag{III.3}$$

where θ is the angle between the axis of the segment and the eventual direction of nematic ordering. Each rod experiences a <u>mean</u> interaction energy

$$\langle U\rangle = -\varepsilon g k_B T \Theta \langle Q\rangle \nu Q \tag{III.4}$$

where g is the volume fraction of the sample occupied by the macromolecules, Θ is the average fraction of monomers in helical segments, and $\langle Q\rangle$ is the orientational order parameter ($\langle Q\rangle = 0$ in the isotropic state; $\langle Q\rangle = 1$ if all the molecules are rigid and completely aligned). The thermodynamic probability that a given rod makes an angle θ with the nematic axis is proportional to

$$e^{h\nu Q(\theta)} \tag{III.5}$$

and the dimensionless mean field h is given by

$$h = \varepsilon g \Theta \langle Q \rangle \tag{III.6}$$

In order to simplify the algebra, let us specialize to the case of sharp helix-coil transitions, i.e., $\Sigma \to 0$ as is common for the synthetic polypeptides. (The general situation will be discussed in a later publication.) For this limit, a given chain is either a flexible coil or a rigid helix, i.e., $\nu = 0$ or N (Fig. 1). The free energy is separable into two contributions: (1) the intra-molecular interactions associated with the helix-coil transition; (2) the intermolecular excluded volume interactions which are responsible for any nematic order. The difference in free energy per chain between a helical and randomly coiled molecule arising from intramolecular interactions is $-N \ell n\, s$. The intermolecular excluded volume interactions in the mean field Meier-Saupe like approximation [Eqs. (III.2) - (III.6) with $\nu = N$] gives a contribution to the free energy difference, $(\Delta F)_{MS}$ which is sketched in Figure 2 as a function of the nematic order parameter $\langle Q \rangle$ for several values of the monomer concentration c. For $s > 1$, the molecules are rigid and a first order lyotropic phase transition is predicted to occur at

$$c^* = N\sigma^* = k(Na^3)^{-1} \tag{III.7}$$

For $s < 1$, a phase transition may only occur if the gain in free energy of ordering exceeds the entropic cost in loss of flexibility of the individual chains. In the (s,c) plane, the phase boundary will then be given by

$$s = \exp[(\Delta F)_{MS}/Nk_BT] \tag{III.8}$$

For $s < 1$, this implies $(\Delta F)_{MS} < 0$ or $c > c^*$. Is there a minimum value of s $(=s_m)$ below which the nematic state is never stable? We may estimate that the minimum value of $(\Delta F)_{MS}$ by setting $\nu_i = N$ and $Q_i = 1$ in (III.2). Then, we find

$$(\Delta F)_{MS} > -\varepsilon k_B TN(ca^3) \tag{III.9}$$

where the factor ca^3 is the volume fraction of the sample occupied by polymer. Thus at the highest concentrations, $ca^3 \to 1$, we obtain

$$(\Delta F)_{MS} > -\varepsilon N k_B T \tag{III.10}$$

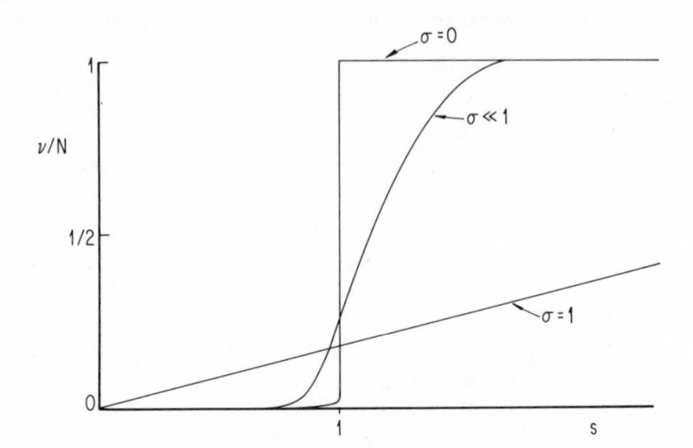

Figure 1. Persistence length ν as a function of s for several values of Σ. For the Ising limit $\Sigma = 1$, the slope of the straight line is N^{-1}. For $\Sigma << 1$, the value $\nu(s = 1) \cong \sigma^{-1/2}$.

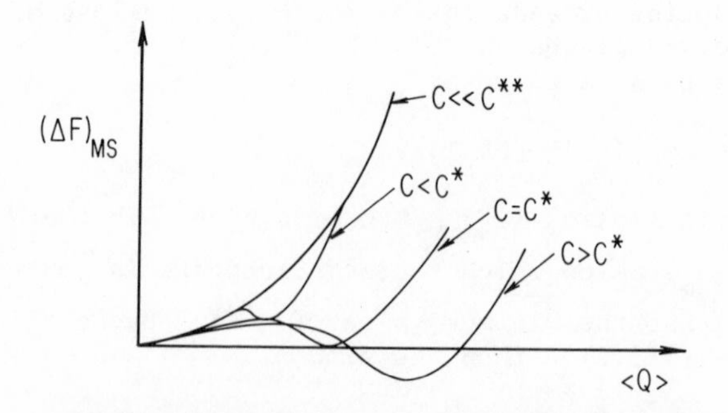

Figure 2. Difference in free energy between nematic and isotropic phases of a rigid rod solution for different values of the monomer concentration c. There is a predicted first-order lyotropic transition at $c = c^* = N\sigma^* = k/Na^3$.

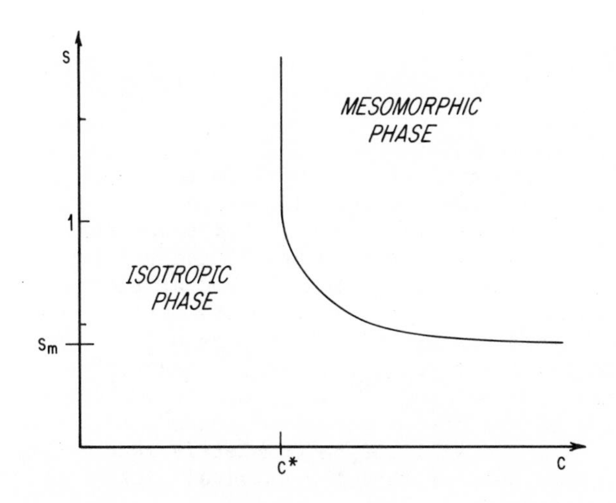

Figure 3. Phase boundary separating the isotropic and nematic phases in the cooperative limit $\Sigma \rightarrow 0$. The area to the right of the solid line and below $s = 1$ represents the region of induced rigidity.

or $s_m \simeq e^{-\varepsilon}$. For $\varepsilon \simeq 1$, $s_m \sim 1/2$, which is a significant renormalization of the helix-coil transition. This phase boundary is sketched in (Fig. 3) for $\Sigma \to 0$. The experimental signature of the induced rigidity part of the boundary would be, for example, jumps in both the rotary dispersion (helix-coil transition) and optical birefringence (isotropic-nematic transition) at the same polymer concentration. Indeed for PBLG in dichloroacetic acid, the early work of Robinson et al.[4] gives some indications of such behavior. More recent studies in the same system by Frenkel et al.[14] and Uematsu[15] provide further evidence in support of the induced rigidity concept.

IV. Final Remarks

The principal goal of this work is to pursue the concept of induced rigidity.[9] Our current investigations include extensions to the cases of arbitrary values of the cooperativity parameter $1 > \Sigma > 0$, more accurate treatments of the rod-rod excluded volume interaction,[16] and the inclusion of some attractive dispersion forces. While we refer to the phase transition as lyotropic, in fact we do expect that it also occurs as a function of temperature (as well as pH, ionic strength, etc.) because of the dependence of s on these parameters. We must further emphasize the semi-quantitative nature of our results. Indeed under most actual circumstances, s and c will not be completely independent variables. Nevertheless we believe that the physical picture presented here has some relation to reality.

Abstract

We have previously suggested[9] that a cooperative helix-coil-nematic phase transition may occur in solutions of heliogenic macromolecules such as the synthetic polypeptides (e.g., PBLG). In this study, we extend the model to the more explicit case of excluded volume interactions only. For the special case of fully cooperative isolated molecule helix-coil transitions, we sketch out the phase boundary between isotropic and nematic regions.

Literature Cited

1. Such systems were discussed in detail at the ACS Symposium on Rigid Chain Polymers, New Orleans, March, 1977, Polymer Preprints (1977) 18, 1, and J. Poly. Sci. (to be published).
2. ACS Symposium on Mesomorphic Order in Polymers, Chicago, August, 1977, Polymer Preprints (1977) 18, 2.
3. Blumstein, A., "Liquid Crystalline Order in Polymers," Academic Press, New York, 1977 (to be published).
4. Robinson, Ward, and Bevers, Discuss Faraday Soc. (1958) 25, 29.

5. Murthy, N. S., I. R. Knox, and E. T. Samulski, J. Chem. Phys. (1976) 65, 4835.
6. Onsager, L., Ann. N. Y. Acad. Sci. (1949) 51, 627.
7. Flory, P. J., Proc. Roy. Soc. (1956) A234, 73.
8. Wadati, M., and A. Isihara, Mol. Cryst. Liq. Cryst. (1976) 17, 95.
9. Pincus, P., and P. G. de Gennes, Polymer Preprints (1977) 18, 131, and J. Poly. Sci. (to be published).
10. The statistical mechanics of helix-coil transitions is well discussed in Birshtein, T. M., and O. B. Pititsyn, "Conformations of Macromolecules," Interscience, New York, 1966, Chapters 9 and 10.
11. Yamakawa, H., and W. H. Stockmeyer, J. Chem. Phys. (1977) 57, 2843.
12. Pincus, P., "Macromolecules" (to be published).
13. See, for example, de Gennes, P. G., "The Physics of Liquid Crystals," Clarence Press, Oxford, 1975, Chapter 2.
14. Frenkel', S. Ya., L. G. Shaltyko, and G. K. Elyashevich, J. Polymer Sci. (1970) C 56, 47.
15. Uematsu, Y., Private Communication.
16. An excellent review of this subject appears in Straley, J. P., Molecular Crystals and Liquid Crystals (1973) 22, 333.

RECEIVED December 8, 1977.

11

Structure and Properties of Polyglutamates in Concentrated Solutions

Y. UEMATSU

Department of Industrial Chemistry, Tokyo Institute of Polytechnics, Atsugi, Kanagawa, 243-02, Japan

I. UEMATSU

Tokyo Institute of Technology, Megro, Tokyo, 152, Japan

It is well known that some polypeptides form lyotropic liquid crystals at sufficiently high concentrarion of the polymers and exhibit the optical properties characteristic of the cholesteric mesophase. The solutions show the regular striation arranged in a finger - printed pattern under a polarization microscope. Robinson et at[1] have reported that the distance between the striation,S, which is the half pitch of the twisted structure, varies with polymer concentration in an inverse manner. They also discussed the molecular arrangement in the twisted structure in the light of X-ray and other measurements. They suggested that the long-range, electric dipole-dipole interactio plays an important role in the formation of the cholesteric structure, and the helical arrangement of dipoles in the polypeptide molecules would tend to impose a unidirectional twist. From the smallness of the twist angle and the fact that it decreases on dilution, they pointed out that the twistangle arises from a dynamic equilibrium.

Samulski et al[2] suggested that an asymetric helical molecular conformation may provide an explanation for the cholesteric structure, and very small oscillations of helices relative to their neighbors in the liquid crystal would be biased to one side of the parallel position because of the chirality of the van der Waals surface of the polypeptide molecules. Further, they found a linear relationship between the pitch and the reciprocal of temperature. This relation holds quite well for a number of thermotropic liquid crystal systems and was explained theoretically by Keating[3] , who treated the macroscopic twist as a rotational analogue of thermal expansion with the dominant anharmonic forces coming from nearest neigh-

0-8412-0419-5/78/47-074-136$05.50/0

bors. The theory gives the following relationship:

$$P = \frac{4\pi dIw}{AkT}$$

where k is the Boltzmann constant, d, the spacing between neighboring planes of molecules, w, the frequency of the excited twist modes, I, the moment of inertia of the molecule, and A, the constant in the cubic anharmonicity term of the anharmonic equation of motion describing the system. However, the pitch for the polypeptide liquid crystals increases with temperature, while it decreases for the thermotropic liquid crystals.

Recently, there are several reports[4] discussing the twist mode of the binary system of the thermotropic liquid crystals composed of a nematic solvent and a cholesteric solute. Another interest is in attempting to pursue the structural similarity between thermotropic liquid crystals and lyotropic one, especially for the dependence of cholesteric pitch on the temperature.

One of the best measures of twist is thought to be the cholesteric pitch. The temperature dependence of cholesteric pitch is then measured for poly-γ-benzyl-L-, and poly-γ-benzyl-D-glutamates (PBLG, and PBDG) and poly-γ-alkyl-L-glutamates, in various solvent systems. The temperature dependence of some physical properties were also measured.

Experimental.

PBLG (mw: 201,000) and PBDG (mw: 197,000) used in this study were suplied by Ajinomoto Co. Ltd. Japan, and some specimens of PBLG were prepared by polymerization of N-carboxyanhydride of γ-benzyl-L-glutamate. Poly-γ-alkyl-L-glutamates (PALG) were prepared from poly-γ-methyl-L-glutamate (mw: 250,000) by alcoholysis by which all methyl radicals in the side chains can be replaced with the desired alkyl radicals. Poly-γ-alkyl-L-glutamates used in this study were with alkyl groups such as ethyl , n-propyl , n-butyl , n-amyl , n-hexyl , n-heptyl, and n-octyl.

From the viscosity measurements in dichloroacetic acid the molecular weights of polyglutamates were estimated. The solutions were prepared by weighing the polymer and solvent in a stoppered bottle. After shaking slowly on a magnetic stirrer for long time, the solutions were put carefully into the cell, which was having parallel sides 2mm apart. Then the top of the capillary of the cell was sealed to prevent the loss of solvent by vaporization. The concentration was calculated by using 0.787 ml/g as the specific volume of PBG, and 0.83 ml/g as that of PALG s, and expressed in the units vol/vol. After on standing, until a regular structure appeared, we measured the half pitch with a polarization microscope. The temperature of the sample was regulated within ± 0.2°C by use of a constant temperature circulator. The measurements were carried out after maintaining the specimens for 12 hrs at each temperature.

Dilatometric measurements were carried out using Hg as a confining liquid, which was transferred into the dilatometer after solidifying the sample solution by liquid nitrogen. CD spectra, were measured with JASCO ORD/UV-50 over the region from 220 to 300nm. NMR spectra were measured by use of JEOL MH-100. The time required to arrive at equilibrium varied from one to several hours. The solvents used here were purified by standard methods.

Results and discussion.

In general the cholesteric pitch of thermotropic liquid crystals decreases with temperature. The theoretical treatments for this behavior have been reported[3,4]. However, it was found that for a binary system, the dependence of the cholesteric pitch on temperature changes from positive to negative at a critical composition. For the binary system of thermotropic liquid crystals, composed of cholesteric and nematic substances, dS/dT is greater than zero in the nematic rich region, and dS/dT is less than zero in the cholesteric rich region. In this paper, we have investigated the temperature dependence of cholesteric pitch for the polypeptide liquid crystals.

Recently DuPre et al[5] reported that, S increases linearly with temperature rise. Qualitatively, their results are consistent with ours. However, the time required to reach the equilibrium pitch, varied with temperature, the concentration of polymer, and also the thermal history. For PBLG solution in dichloroethane (EDC), which concentration is 0.12 vol/vol, the variation of pitch with time was measured at a constant temperature by T-jump method. Fig. 1 shows the time dependence of cholesteric pitch by the T-jump method from -2°C to +30°C, 40°C and 50°C respectively. It is clear that the time required to arrive at the equilibrium pitch is shorter at higher temperature but is still over several hours. Therefore, the equilibrium pitch must be measured after prolonged aging at each measuring temperature. It was found that the half pitch, S, increased monotonically with temperature, and finally, retardation line disappeared at a critical temperature, and the birefringence characteristic of the nematic structure was observed.

Plot of S versus temperature for PBDG in chloroform is shown in fig.2. This behavior corresponds to the behavior of binary mixture of thermotropic liquid crystals in the nematic rich region. As shown in fig. 3, the reciprocal of the half pitch which is proportional to the angle of twist between $(10\bar{1}0)$ planes, is a linear function of temperature. The critical temperature, where $1/S = 0$, was obtained from the extrapolation of this curve, and was defined as the nematic temperature, T_N. The optical pattern characteristic of the nematic state was observed in this region under the polarization microscope. DuPre et al defined this temperature as the 'clearing point', since the twist elastic constant k_{22}, obtained from the orientation in a magnetic field tends toward zero at this temperature. However, from the observation by a polarization microscope, it is obvious that the newly appeared phase is not an isotropic one. The fact that k_{22} tends toward zero, can be explained with the presence of the size of the cluster of coherent length, in which the molecules would be arranged parallel.

In order to obtain the relationship between S and the concentration C, a double logarithmic plot of S against C at each temperature was made as shown in fig. 4. It can be seen that log S is proportional to log C and, its slope n is an increasing function of temperature, characteristic of the specimens used. The extrapolated line of log S-log C crossed each other at a critical concentration C_0 at which S stays constant and independent of temperature. These results suggest that the temperature dependence of the cholesteric pitch would inflect at the concentration higher than C_0. This is analogous to the behavior of thermotropic liquid crystals composed of cholesteric solute and nematic solvent, where the sign of dS/dT reverses at a critical concentration. It is understood that the behavior of both thermotropic and lyotropic liquid crystals is comparable provided that the nematic substances of the former are substituted with the solvents of the latter. The critical concentration C_0 is about 0.41 vol/vol and this value is very close to the concentration at which the side chains on neighboring molecules of the polymer come to contact each other (refer to fig.5). From these results, it is expected that the origin or mechanism of twist would change at this concentration C_0. The

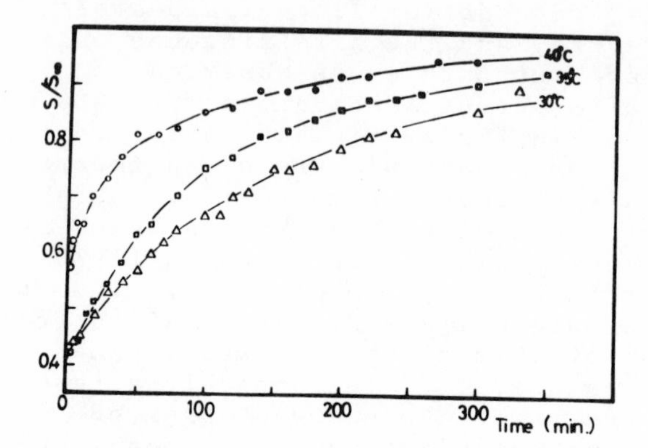

Figure 1. The change of cholesteric pitch with time for PBLG–EDC. Polymer concentration is 0.12 vol/vol.

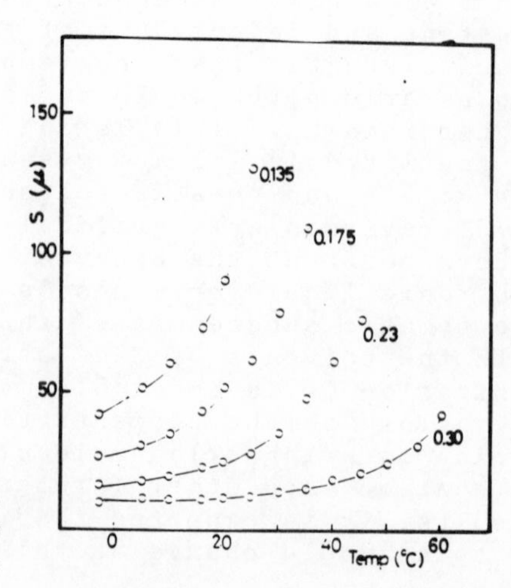

Figure 2. The change of S with temperature for PBDG–$CHCl_3$. Numbers represent the concentration of polymer.

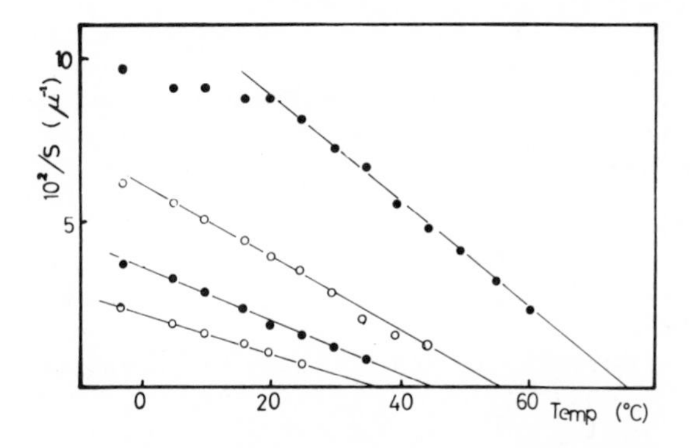

Figure 3. 1/S vs. temperature for PBDG–EDC. Concentration (from top to bottom) 0.30, 0.23, 0.175, 0.135 (vol/vol).

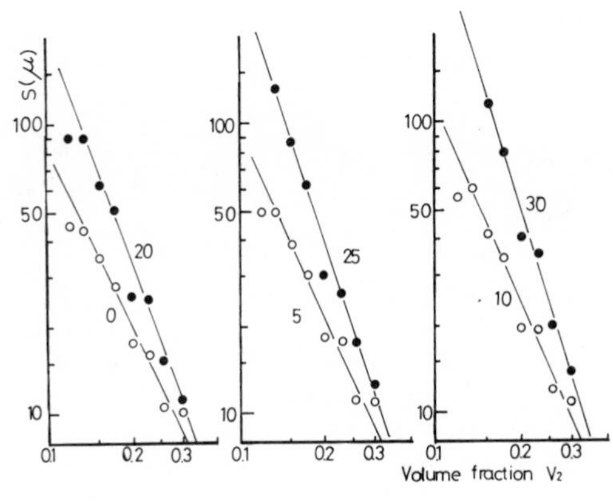

Figure 4. Log S–log C plot for PBDG–$CHCl_3$ at each temperature (°C)

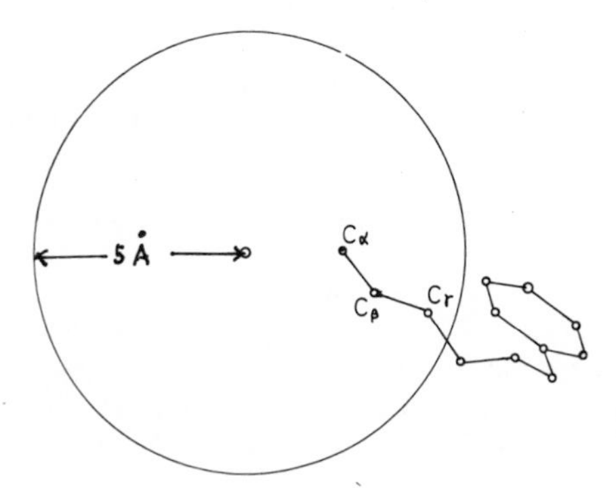

Journal of Molecular Biology

Figure 5. PBLG conformation (projection along the helix axis) (7)

extrapolation of the plot of log S versus log C, to C = 1, gives the half pitch of pure polymer, which is a decreasing function of temperature; and this behavior is analogous to that of thermotropic cholesteric liquid crystals.

It is convenient to use the reduced temperature T/T_N, where T_N is the nematic temperature, instead of temperature T, in the temperature range near T_N. Fig. 6 shows the observed values of 1/S versus T/T_N, for PBDG in chloroform, at various concentrations. The slope B depends on the concentration, and it increases with concentration. If the cholesteric pitch was measured at constant concentration, the following equation holds:

$$1/S = B\,(\,1 - T/T_N\,)$$

The dependence of B on the concentration of polymer was obtained from the plot of log 1/P or 1/S against log C, at each T/T_N. The values of the slope n are constant and independent of temperature(fig. 7).

According to Robinson[1], the following relation holds:

$$\theta = 2\pi d/P \text{ or } \pi d/S \text{ and } C \propto 1/d^2$$

Hence

$$\theta = (\,A/d^{2n-1})(\,1 - T/T_N)$$

where d is the distance between $(10\bar{1}0)$ planes, and 2n-1 varies from 2 to 6. The fact that the twisting angle θ was inversely proportional to d^{2n-1} would suggest that the twisting power originates from a long range repulsive force between the molecules, which might be an electrostatic repulsive force or dipole-dipole interaction. The values of A are calculated from the slope of the plot of 1/S versus T/T_N, and C^n, at each concentration. The obtained values of A for PBLG in various solvents are listed in table 1. As shown in the table, these values are clearly constant and independent of concentration.

In order to study the mechanism of the transition of cholesteric to nematic structure, several physical properties were measured as a function of temperature from room temperature through this transition region.

<u>Specific volume measurement</u>. The specific volumes of PBLG solution in chloroform, dichloromethane, dichloroethane and dioxane were measured respectively by dilatometry. It can be seen, from fig. 8 that the expansion coefficient varies abruptly at T_N. These

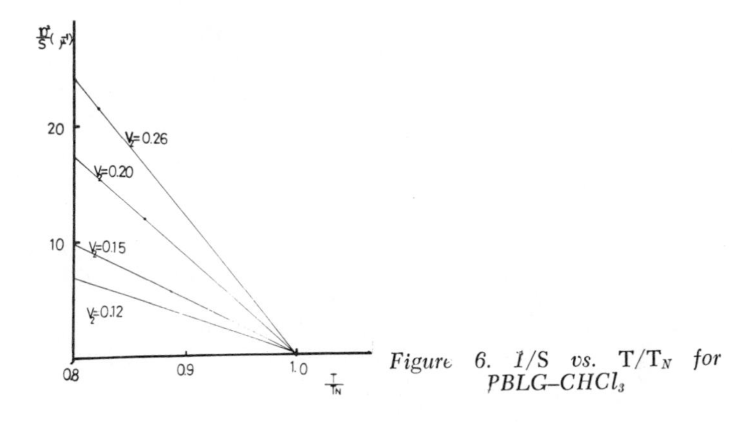

Figure 6. 1/S vs. T/T_N for PBLG–$CHCl_3$

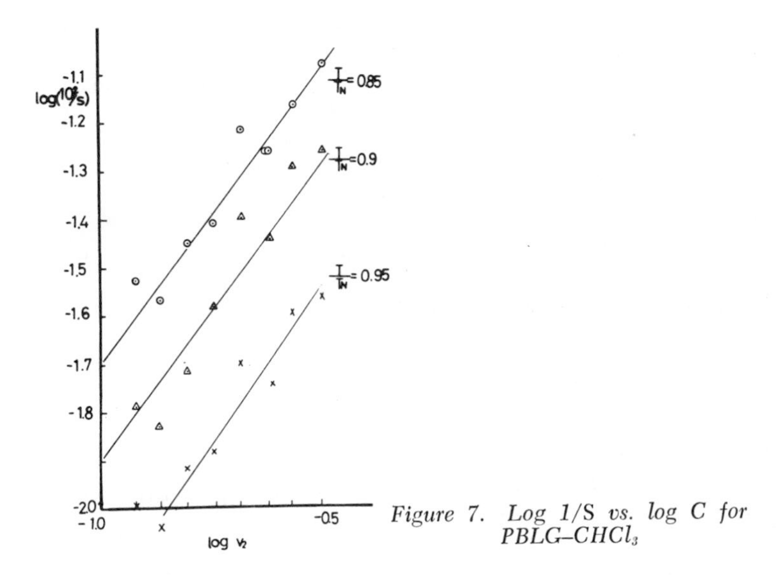

Figure 7. Log 1/S vs. log C for PBLG–$CHCl_3$

Table I The values of constant A

	n \ C*	0.12	0.14	0.16	0.18	0.20	0.23	0.26	0.30	ave.
$CHCl_3$	1.42	4.08	3.10	3.42	3.09	3.90	2.90	3.45	3.07	3.5
CH_2Cl_2	1.58	6.71		5.96		6.32				6.3
CH_2ClCH_2Cl	1.42	7.10		7.00		7.50		8.20		7.5
$CHCl_2CHCl_2$	0.86	3.47		4.50		4.43		4.46		4.46
$CH_2ClCHClCH_2Cl$	2.04	25.7		32.7		31.7		33.7		32.7

* The concentration of PBLG represented in vol/vol

results suggest that the cholesteric to nematic transition may be a second order transition and the occurence of some molecular motion should be expected. This is also supported by the fact that a λ-like endothermic peak was observed near the nematic temperature in the differential scanning calorimetry.

Circular dichroism. CD spectrum for PBLG in ethylene dichloride was measured. The induced CD band by π-π* transition of phenyl ring in the side chain appeared, centered at 260nm. It is clear that the band is induced by cholesteric structure, since its intensity decreases with the orientation by the magnetic field and finally tends to zero. We measured the temperature dependence of the intensity of this band, which decreased monotonically with increase in temperature, but as against our expectations, it remained within an insignificant value above the nematic temperature. The result implies that the chiral structure on the side chains is partially preserved even above the nematic temperature. The difference between the nematic structures induced thermally and magnetically might be caused by the difference of the arrangement of the side chains.(fig.9)

Nuclear magnetic resonance. Sobajima[6] has reported that the magnetic field induced a change in the cholesteric structure of polypeptide liquid crystal, and the NMR spectrum of CH_2Cl_2 proton in PBLG-dichloromethane changes to doublet because the direct dipole-dipole hyperfine interactions are not averaged to zero. They derived an equation for the dependence of the equilibrium splitting with temperature from the Gutowsky and Pake equation, as given below, assuming that the solvent molecules are oriented according to a Langevin mode in the electric dipolar field generated by polymer molecules.

$$h = \frac{\mu o}{10\ r^3}(\ 1 - 3\ \cos^2\theta\ )\ (\ \frac{\mu E}{kT}\ )^2$$

where h is the splitting width, μo the magnetic dipolemoment of proton, r,the interproton distance, θ,the angle between polymer orientation axis and magnetic field, μ,the electric dipole moment of solvent molecule, E,the electric dipolar field by polymer molecules, k,Boltzmann constant, and T,the absolute temperature.

The plot of the splitting width h versus $1/T^2$ is shown in fig.10, in which the broken line shows the

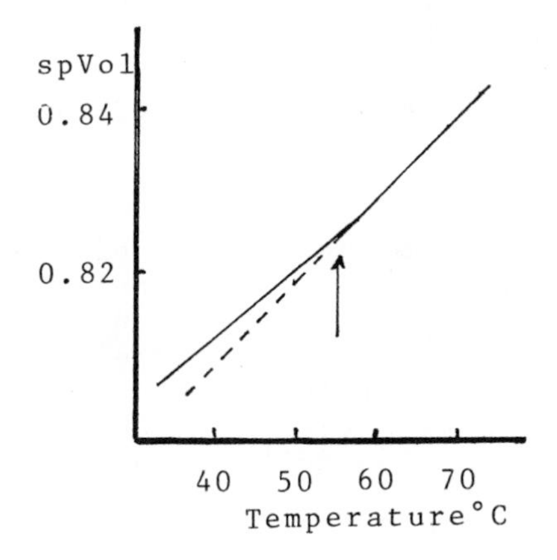

Figure 8. Specific volume vs. temperature for PBLG–EDC (C = 0.15)

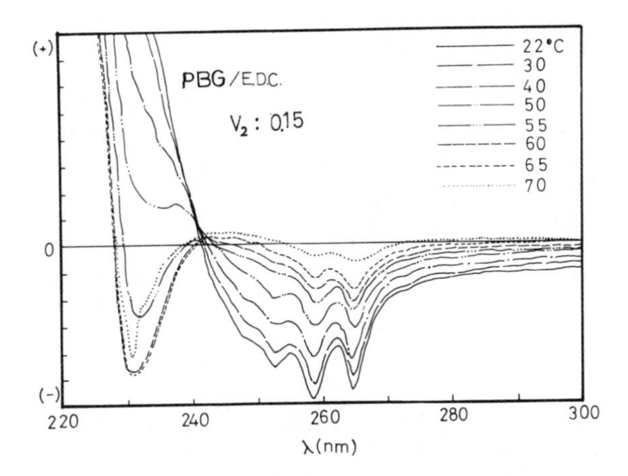

Figure 9. CD spectrum for PBLG–EDC

calculated values, and deviation from the theoretical curve starts at the temperature 10° lower than the nematic temperature. Furthermore, the observed values are smaller than the theoretical value. The deviation would be explained by the increase of the free solvent molecules or weakening the dipolar field generated by polymer molecules,by the motion of side chain.

Fig. 11 shows the concentration dependence of splitting,h, which increases with polymer concentration, that is with the decrease of the distance between dipoles of neighboring molecules. In rather dilute solution, the NMR spectrum of CH_2Cl_2 proton in PBLG-EDC-CH_2Cl_2 mixture was measured with different compositions. The spectrum changed from doublet to triplet with the addition of EDC, and the intensity of the center peak increases with EDC fraction increase. The equilibrium separation of outer peaks is increasing slightly and their intensity decreases remarkably with increase of EDC fraction. The fact that the decrease of the intensity of outer peaks occurs in the range of 0.4 to 0.6 volume fraction of EDC, suggest that CH_2Cl_2 molecules binding to side chains are outwards and EDC molecules might have been substituted. This result implies that the change of h can not be explained by the increase of free molecules. At the higher polymer concentration, the temperature dependence of h of CH_2Cl_2 protons in mixed solvent system was measured. As shown in fig.13, the center peak does not appear in this case different from the result shown in fig.12. The results are summerized in table 2. The smallness of h of CH_2Cl_2 proton in benzene and dioxane are explained by assumption that benzene and dioxane molecules might bind with ester groups in the side chains closer than CH_2Cl_2 molecules or they might randomize the arrangement of the side chains. However, it is thought, from the fact that benzene and dioxane promote gelation as well as from the information of side-chain conformation in the film cast from corresponding solvents, that they make side chain fixed. Thus the former mechanism may be supported. However, the fact that the center peak was not found in this system, suggests that CH_2Cl_2 molecules are still located near the side chain and get oriented in the electric dipolar field.

<u>The effect of solvent on T_N and twisting angle</u>.
It was found that the cholesteric pitch appeares again at higher temperature, and chilarity at higher temperature is opposed to that at lower temperature region, as shown in fig.14. The sign of ORD of films cast

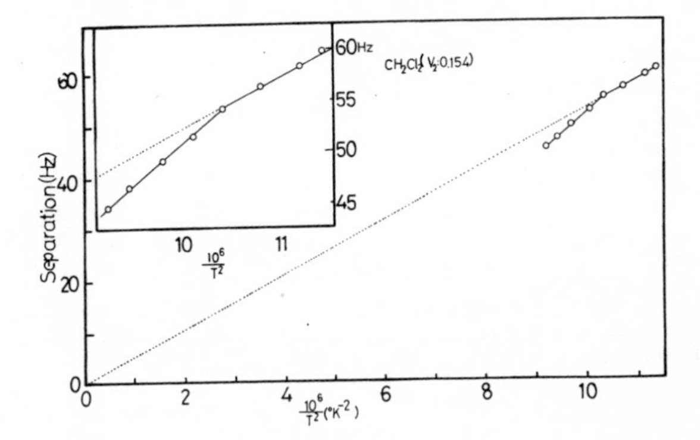

Figure 10. Separation of CH_2Cl_2 proton in PBLG–CH_2Cl_2 vs. $1/T^2$ where the concentration C *is 0.154*

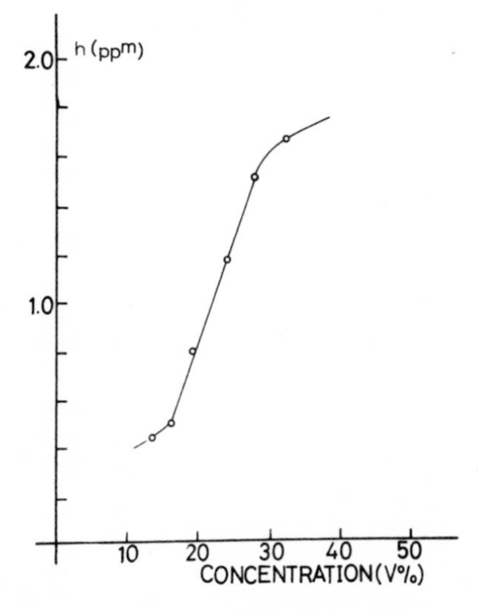

Figure 11. The change of h *of CH_2Cl_2 with polymer concentration (PBLG)*

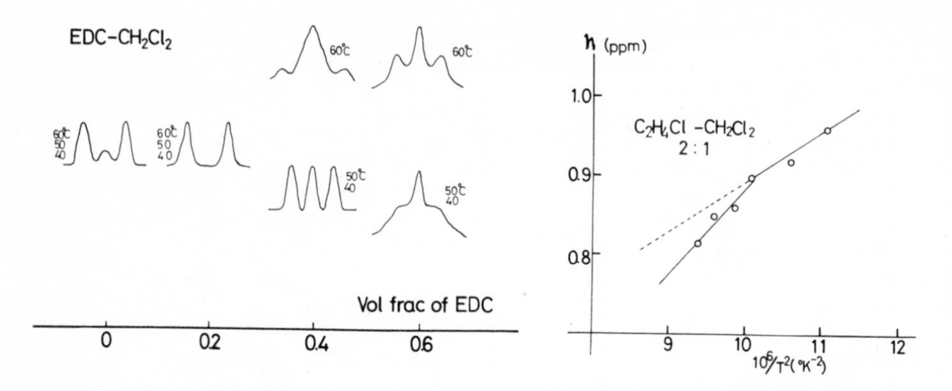

Figure 12. (a) (left) *NMR signal of CH_2Cl_2 for PBLG–EDC–CH_2Cl_2 (C = 0.19 vol/vol) (b)* (right) h *vs. 1/T^2 for PBLG–EDC–CH_2Cl_2* (C = *0.19)*

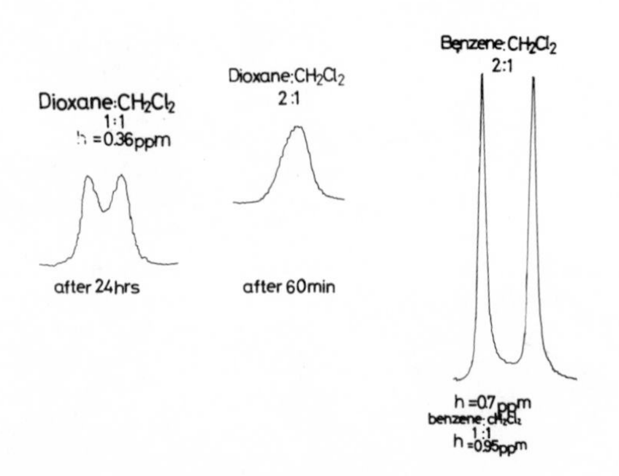

Figure 13. NMR signal of CH_2Cl_2 for PBLG–CH_2Cl_2–dioxane and PBLG–CH_2Cl_2–benzene (C = *0.19)*

Table II The splitting of CH_2Cl_2 proton at 30°C. The concentration of PBLG: 0.20 vol/vol

	$CHCl_3$	EDC	benzen	dioxane
h_{ppm}	0.8	0.98	0.7	0

Solvent compositions in binary solvent systems are 1/3 of CH_2Cl_2.

under the corresponding conditions is opposite to each other. In lyotropic liquid crystal of PBLG or other polypeptides, since the size of solute molecules and solvent molecules are quite different, the latter can move more easely than the former and the solvent molecules are forced to orient themselves between solute molecules. Thus it seems that the local intermolecular interactions and the molecular arrangement of solvents might be of importance in determining the helical pitch and twisting direction. The local intermolecular interactions of polymer-solvent indicate the specific shapes and sizes of the individual solvent molecules and their orientation between long rigid rods.

The sense of cholesteric twist, in the liquid crystal of PBLG in dioxane, is opposed to that in CH_2Cl_2, as pointed out by Robinson. We measured the cholesteric pitch of PBLG in various mixed solvent systems, and estimated the sense of cholesteric twist in the individual solvent. If two solvents which make the sense of cholesteric twist of PBLG opposite to each other, are mixed, the cholesteric pitch of PBLG in mixed solvent will diverge at the critical composition. It is found that the sense of cholesteric twist of PBLG in dioxane and chloroform is opposite to that in dichloromethane, dichloroethane and benzene.

The dependence of nematic temperature on the size of solvents is illustrated in fig. 15. The largest interatomic distances of the solvent molecules are taken as their sizes. The larger the solvent molecule, the higher is the nematic temperature in general. Further, the solvents which have dipole moment parallel to the long axis depress nematic temperature and others which have dipole moment perpendicular to it, elevate T_N.

We attempted to offer the structural model in cholesteric molecular arrangement. By now, it is well known that the PBLG molecule has a large dipole moment parallel to the helical axis, and on the other hand, the dipoles of side chain are arranged inversely to that of main chain and phenyl group is oriented with its three principle axes at 46°, 57° and 62° with respect to the fiber axis respectively and the phenyl group cannot be horizontal nor vertical, but must be inclined. The stable arrangement between neighboring molecules may be taken as a rough model to minimize the steric hindrance between side chains, as shown in fig. 16. Furthermore, this arrangement also makes the electrostatic energy to be at a minimum. The

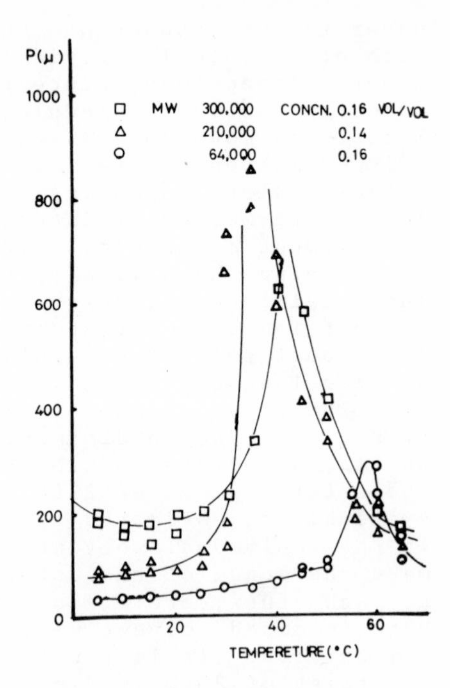

Figure 14. p vs. temperature for PBLG–EDC

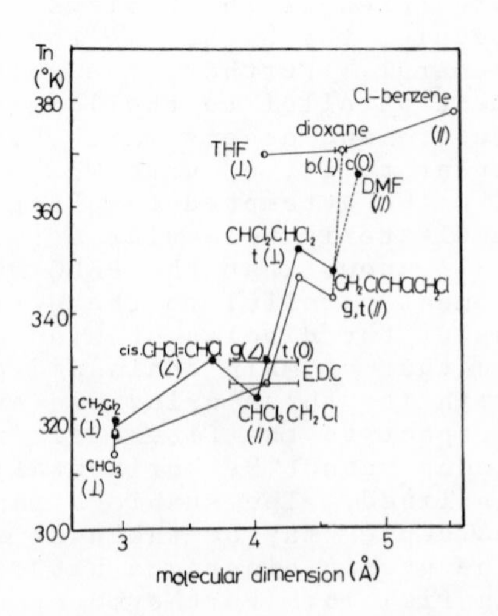

Figure 15. Change of T_N with size of solvent molecule

most plausible arrangement is shown schematically in fig. 17, in which the equal numbers of molecules having dipoles of opposite direction, are contained and the maximum numbers of antiparallel pair are also contained.

The twisting force between antiparalleled molecules should be smaller than that of paralleled molecules. Then the Z-axis should be the cholesteric axis in the isotropic hexagonal array, keeping the hexagonal arrangement locally. This model agrees with the results of X-ray diffraction analysis by Robinson. It is also suggested that the twisting power should be a long range repulsive interaction, and the asymetric array of permanent dipoles on side chains of a polymer molecule should impose a unidirectional twist on the nearly parallel helices.

The effect of side-chain length. As shown in fig. 18, the behavior of poly-γ-butyl-L-glutamate (PBuLG) in $CHCl_3$ represents an analogous behavior to PBLG, i.e. 1/S decreases linearly with temperature. However, in the solution of poly-γ-amyl-L-glutamate (PAmLG) in $CHCl_3$, it is observed that, at first, 1/S decreases linearly with temperature, goes through zero, namely where there is a nematic state and the cholesteric pitch disappeared; and thus 1/S begins to increase, as shown in fig. 19. It is found from ORD measurement that the sign of cholesteric twist changes at T_N. In the $CHCl_3$ solutions of poly-γ-hexyl-L-glutamate (PHexLG) which have longer alkyl group in the side chain, 1/S increases with temperature, as shown in fig. 20. The behaviors of poly-γ-heptyl-L-glutamate (PHepLG) and poly-γ-octyl-L-glutamate (POLG) in $CHCl_3$ are analogous to that of PHexG. This suggests that its nematic temperature is lower than the room temperature. The shorter is the side chain, the higher is the T_N, generally in $CHCl_3$. Then, the polymers which have the longer side chains show a different sign of cholesteric twist, from that which have the shorter side chain, under the same condition.

As stated already, the sign of cholesteric twist of the solution of PBLG in $CHCl_3$ is opposite to that in dichloroethane. The cholesteric twist of PALG also behaves in the same manner, that is, the sign is different to each other in the two solvents. However, the nematic temperature of PALG in EDC increases with the increase of alkyl chain length in the side chain. The constant n, then can be obtained from the plot of log 1/S against log C in EDC. It

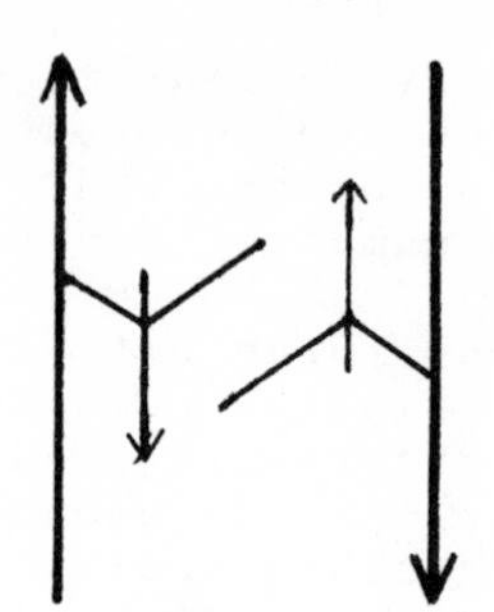

Figure 16

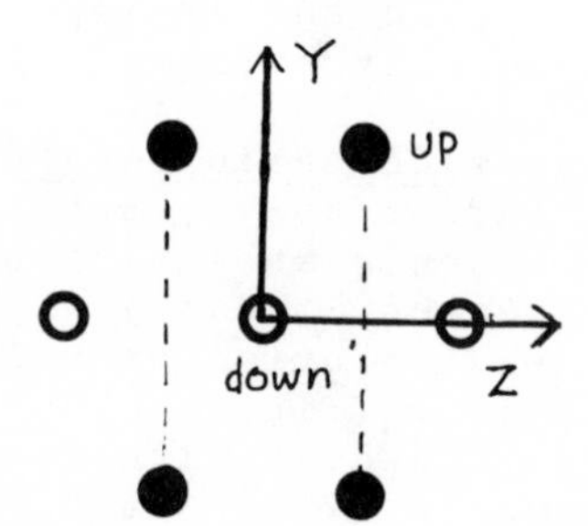

Figure 17

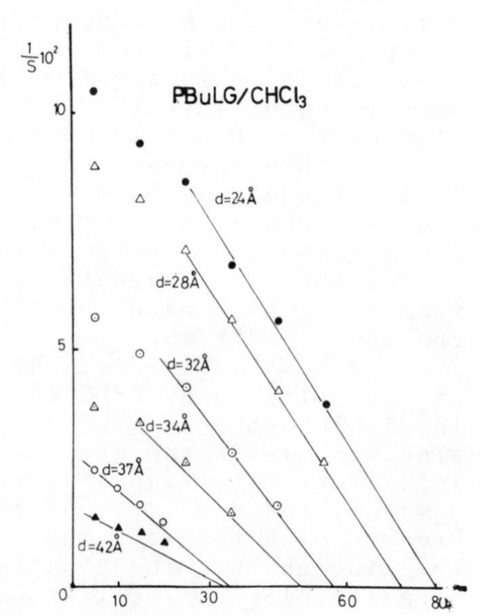

Figure 18. 1/S vs. temperature for PBuLG–$CHCl_3$

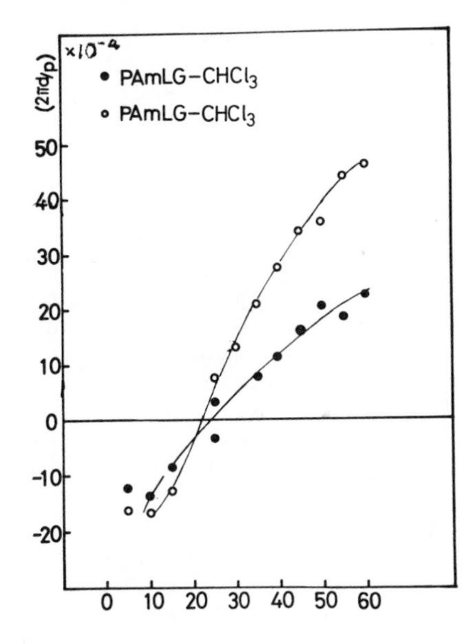

Figure 19. Twist angle vs. temperature. (●) C = *0.14*; (○) C = *0.25.*

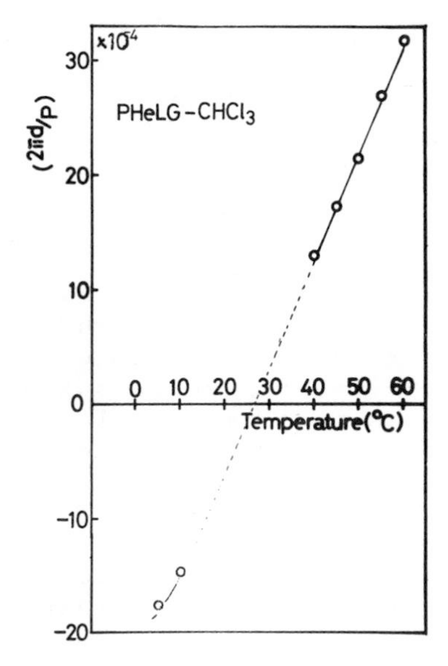

Figure 20. Twist angle vs. temperature

increases with side chain length. This fact implies that the short range interaction will overcome the long range interaction (for example, dipole-dipole interaction) with the increase of alkyl chain length in the side chain.

As shown in fig. 21, d/S, which is directly proportional to twisting angle, increases with decrease of d. However, the value of d/S goes through the maximum and begins to decrease at a critical d. The vertical lines indicated in fig. 21 , represent d at which the side chains begin to come in contact, assuming fully extended conformation of the side chains. The interaction between side chains may lead to decrease of twisting angle and force polymer molecules to transform to another stable arrangement.

The distance between dipoles of neighboring molecules increases with the increase of side-chain length, and then, the strength of dipole-dipole interaction between neighboring molecules weakens. These results suggest that the twisting direction is affected not only by long range interaction (dipole-dipole interaction), but also by short range interaction (van der Waals repulsive force),with the increase of side chain length, and the twisting power may be affected, too.

The difference between PBLG and PALG may be in their side-chain conformation. As already pointed out, the side chain of PBLG is not fully extended. On the other hand, in PALG the side chain might be extended more than that in PBLG. This fact is also supported by the experimental results, that is, the motion of side chain of PALG, even in solid state, begins at a temperature lower than that of PBLG. Besides, the longer the side chain, the lower the transition temperature to begin the motion of side chain exhibited.

Conclusion.

The cholesteric structures for polypeptide liquid crystals are very complicated, such as those which exhibit inversion of the screw sense with temperature, solvents and side-chain length. Our results suggest that the twisting force between paralleled molecules should be larger than that of antiparalleled molecules and asymetric array of permanent dipoles in side chains should impose a unidirectional twist between neighboring molecules. However, the arrangement and motion of side chains, that is, local intermolecular interaction, play an important role in determining the structure of cholesteric helix.

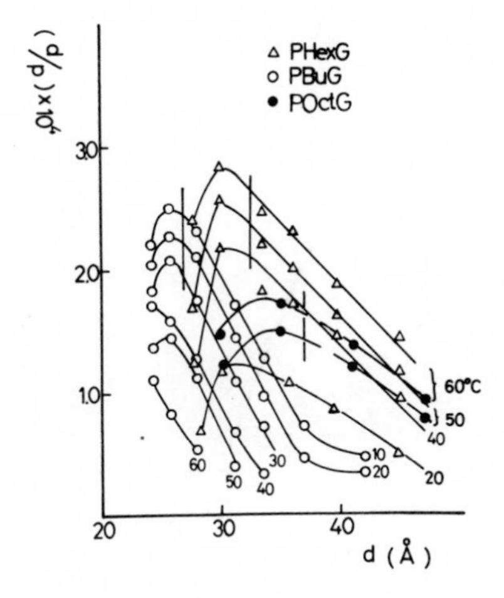

Figure 21. Twist angle vs. d *for PALG–$CHCl_3$*

1) C.Robinson; Mol.Crystals, 1 467 (1966)and references cited therein.
2) E.T.Samulski and A.V.Tobolsky; Liquid Crystals and Plastic Crystals, vol.1,Ellis Horwood Pub.(1974)
3) P.N.Keating; Mol.Crys.Liquid Crys. 8,315 (1969)
4) for example, H.Hanson et al; J.Chem.Phys.,62 1941, (1975), A.Wulf; J.Chem.Phys.60,3994 (1974), Y.R. Lin-Lin et al; Phys. Rev.,14, 445 (1976)
5) D.B.DuPre and R.W.Duke; J.Chem.Phys.,63,143 (1975)
6) S.Sobajima; J.Phys.Soc.Japan, 23, 1070 (1967)
7) Y.Mitsui et al; J.Mol.Biol.,24, 15 (1967)

Received March 13, 1978.

12

Magnetic Reorientation and Counterrotation in Poly(γ-Benzyl Glutamate) Liquid Crystals

ROBERT W. FILAS

Department of Chemistry, Princeton University, Princeton, NJ 08540

Poly(γ-benzyl glutamate) (PBG) is a synthetic polypeptide which adopts the α-helical conformation in various organic solvents. Its essentially rod-like shape is responsible for the formation of a liquid crystalline phase above a critical concentration of polymer (1,2). The nature of this mesophase is usually cholesteric (2,3) due to the chirality of the PBG molecules, but particular solvent mixtures (4) or a racemic mixture of the D and L enantiomorphs (3,5) form a nematic phase. NMR studies have shown that in a magnetic field the PBG molecules tend to align parallel to the field in a nematic-like structure (6-9). If such a magnetically oriented sample is rotated by some angle, θ_0, the reorientation process can be described very accurately (10), but above a critical angle the reorientation mechanism becomes more complicated. The purpose of this paper is to report the detection of this critical angle by NMR, optical, and viscometric techniques in solutions of PBG in dichloromethane.

Experimental

Liquid crystalline solutions of PBDG were prepared using reagent grade dichloromethane and sealed in NMR tubes. Their concentrations were determined gravimetrically and are expressed in w/w percent. Each sample contained a 0.38 mm diameter stainless steel sphere used for viscosity measurements. The racemic mixture, abbreviated as (D+L)PBG, is the same sample used in a previous study (10), and is composed of equal masses of PBLG and PBDG having molecular weights 270 000 and 217 000, respectively.

NMR spectra were recorded on a Varian HA-100 spectrometer in an "unlocked" mode using an external oscillator and frequency counter to calibrate its sweep parameters. Samples were equilibrated in the magnetic field without spinning at ambient temperature (*ca*. 32°C) for nearly a day before each reorientation. The temperature was controlled with a precision of ±0.2° with a Varian temperature-control unit. The samples were rotated by accurately known amounts with the aid of small aluminum sleeves

0-8412-0419-5/78/47-074-157$05.00/0

attached to the tube holder.

Apparent viscosities were determined using a falling sphere method on samples that had been matured for more than a year. As in the NMR experiments, the samples were magnetically oriented with the field direction perpendicular to the NMR tube axis. Crossed polarizers were mounted on the magnet for correlation of optical and viscometric data. After a rotation experiment was performed, the sample was removed from the magnetic field and placed in a $25.0 \pm 0.02^{\circ}C$ constant temperature bath. The velocity of the sphere falling along the axis of the NMR tube, measured using a cathetometer and timer, was used to calculate the Stokes' law apparent viscosity. A Faxen correction (11) of about 5% was applied to all data.

Results and Discussion

The equations of motion for a memory-dependent nematic liquid undergoing reorientation in a magnetic field have recently been presented (10). The equations were derived using the theory of micropolar continuum mechanics as introduced by Eringen (12, 13). In the special case when the memory can be neglected, the result is

$$\theta(t) = \tan^{-1}(\tan\theta_0\, e^{-At}) \qquad (1)$$

where $\theta(t)$ is the instantaneous orientation of the microelement, and θ_0 is the value of θ at $t = 0$. The parameter A is defined by $A \equiv \chi_a H^2/C$, where χ_a is the anisotropy of the diamagnetic susceptibility, H is the magnetic field strength, and C is the apparent rotational viscosity coefficient. A convenient method for obtaining $\theta(t)$ data in the present case is to monitor the time dependence of the NMR signal of the solvent. The dipolar coupling of the proton pair on each CH_2Cl_2 molecule produces a doublet whose separation (ΔH) varies with the orientation of the surrounding PBG helices. In terms of the equilibrium separation, ΔH_0, the relevent expression is

$$\Delta H(t) = \frac{\Delta H_0}{2}\,(3\cos^2\theta(t) - 1). \qquad (2)$$

The validity of Equation 2 relies upon the assumption that the influence of the anisotropic environment of the polymer molecules on the solvent remains constant during a reorientation. It is therefore implied that the PBG helicies retain their original degree of parallelism on a scale which is large compared to the distance a solvent molecule diffuses during its spin lifetime.

Using the technique outlined above, Equation 1 has been tested under a variety of conditions. The results indicate that Equation 1 is only capable of describing the reorientation when θ_0 is less than some critical angle, θ_c, which varies with the sample and the conditions of the experiment. For example, Figure 1 displays $\theta(t)$ data for a racemic PBG sample and the corresponding theoretical curves generated by Equation 1 using $A = 0.0203\ min^{-1}$. The agreement between theory and experiment is excellent for small values of θ_o (solid dots), but when $\theta_o = 44.4^o$ (open squares), the reorientation appears to proceed more rapidly than predicted by the theory. Even by increasing A to 0.0237 min^{-1}, agreement with Equation 1 can only be obtained for about the first 14 minutes. For slightly larger values of θ_o (open triangles), the deviation is much more pronounced and the data can only be approximated by Equation 1 for about 4 minutes if A is chosen to be 0.0321 min^{-1}. In addition, if several consecutive reorientations are performed with θ_o just above θ_c, the rate of reorientation appears to increase in each successive experiment. This apparent change in the physical properties of the system, or "yield" behavior, is small very close to θ_c, but rapidly becomes more significant as θ_o increases. When θ_o is a few degrees above θ_c, the "yielding" is often quite abrupt in racemic PBG samples, as shown in Figure 1. For values of $\theta_o > 51^o$, no part of the reorientation curve could be fit by Equation 1.

The deviation from the predicted reorientation curves described above suggests that some fundamental change in the reorientation mechanism might be occurring. This conjecture is supported by the changes in the appearance of the NMR spectra during "yielding". The pair of peaks begin to shorten and broaden, with shoulders or small additional peaks appearing between them, indicating a disruption of the original degree of order of the PBG molecules. None of these changes in the NMR line shape occurs when $\theta_o < \theta_c$. Even after several consecutive reorientations below θ_c, the same "A" value is found to apply for each reorientation and the only variation in peak height as the peak separation changes is due to magnetic susceptibility effects (15).

The disruption of the basic structure of the liquid crystal, for any reason, can have serious implications concerning the use of Equations 1 and 2. For example, if the "microelement" of the continuum theory is composed of a collection of PBG molecules acting in a cooperative fashion, then any change in the PBG degree of order contradicts the assumption of a constant microelement, upon which the derivation of Equation 1 is based. The conditions requisite for the use of Equation 2 are also violated once disruption begins to occur. The values of θ plotted in Figure 1 with □ and △ are, therefore, not to be taken literally as angles, but rather as "apparent" angles.

The reason for the existence of a critical angle and the disruption that occurs when $\theta_o > \theta_c$ can be explained by the fact that even at equilibrium with the field not all the PBG molecules are perfectly aligned. Orwoll and Vold (14) have shown, for example, that only about 87% of the polypeptide helicies in their sample (17.5% PBLG/CH_2Cl_2; molecular weight 310 000; 14.1 kG field) were within 20° of the field direction. If θ_o is large enough so that some PBG molecules are rotated beyond 90° to the field, then some fraction of them will reorient in the opposite direction from the remainder. This process has been called "counterrotation" and is responsible for at least a partial randomization of the PBG axes (14). Figure 2 shows a sequence of spectra of a 27.2% PBDG sample at various times after a 75° rotation of the sample. In less than half a minute the original doublet (A) has split into a quartet (B), the inner peaks approaching each other while the outer two are separating. In (C), the two inner peaks have merged into one, and in (D) they have separated again, now moving apart. As the reorientation continues (E-I) the peaks first broaden, then sharpen again, eventually returning to their original shape and separation as in (A). These results are very similar to those presented by Orwoll and Vold and are good evidence for the existence of counterrotating regions.

It is also possible to detect the onset of changes in the physical properties of the system by other procedures. For example, the apparent viscosity (η) of the liquid crystal, as measured by the falling sphere method, is very sensitive to any disruption of the orientation of the PBG molecules. When $\theta_o < \theta_c$, η is found to remain constant at all times during the reorientation. Above θ_c, however, η first decreases, goes through a minimum, then increases back to its original value. The maximum amount of counterrotation (and disruption) occurs when $\theta_o = 90^o$, and the change in η as a function of time after such a reorientation is shown in Figure 3. Of course, the viscosity of the liquid crystal can only be sampled once during any given experiment, so each data point represents a separate sequence of orientation, rotation, and measurement.

In order to accurately determine θ_c from viscosity measurements, it is desirable to amplify the small disruptive effects that occur when $\theta_o \gtrsim \theta_c$. This is easily accomplished by performing several consecutive reorientations. Figure 4 shows the apparent viscosity of a 24.5% PBDG/CH_2Cl_2 sample subsequent to three consecutive reorientations of θ_o, each for five minutes. Using this procedure, the concentration dependence of θ_c was determined at 25^oC and is shown in Figure 5. The variation in θ_c is linear within experimental error and decreases from about 50° to 42.5^o between 9% and 27%. Since the concentration at which the side chains of PBG molecules begin to overlap (based on Robinson's data (3)) is approximately 27%, an interesting future study would be the determination of θ_c for more concentrated samples in order to study the effect of side chain interaction on θ_c.

Changes in the optical properties of the sample are also noticeable when disruption occurs, and by observing the sample between crossed polarizers, θ_c can be determined by the appearance of birefringence colors. These colors appear most rapidly and are most intense when $\theta_o = 90^o$. From a comparison of photographs taken of a 90° reorientation observed between crossed

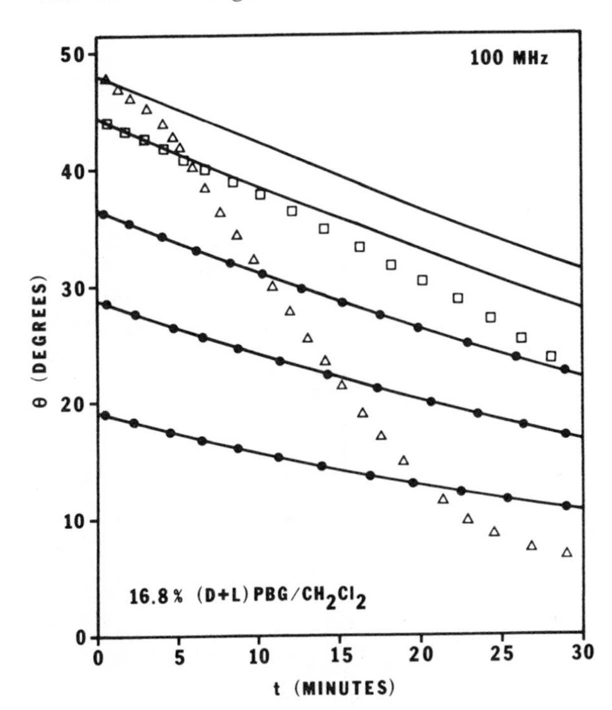

Figure 1. Several 100-MHz reorientation curves for a 16.8% (D + L) PBG/CH_2Cl_2 sample at 2.4°C. (——), Equation 1 with A = 0.0203 min^{-1}. (●) θ_o = 19.2°, 28.8°, 36.7°; (□) θ_o = 44.4°; (△) θ_o = 48.1°.

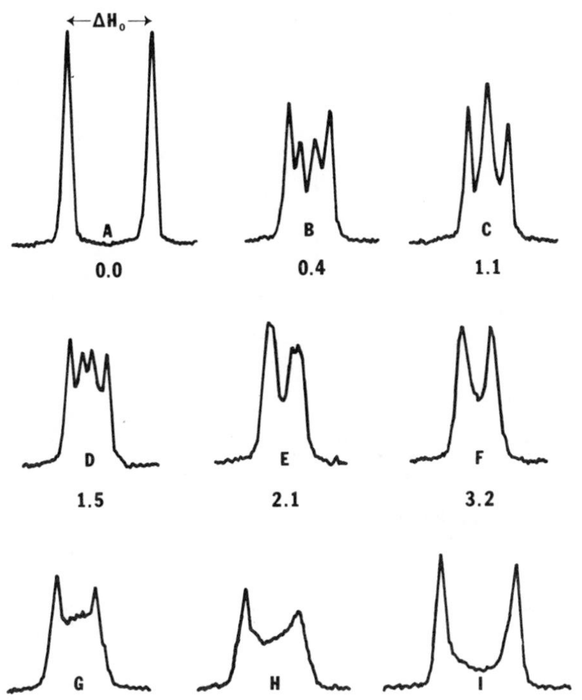

Figure 2. 100-MHz NMR spectra of equilibrium (A) and intermediate (B-I) line shapes subsequent to a 75° rotation of a 27.2% $PBDG/CH_2Cl_2$ sample at −15.1°C. Numbers indicate time after rotation in minutes. ΔH = 200.1 Hz.

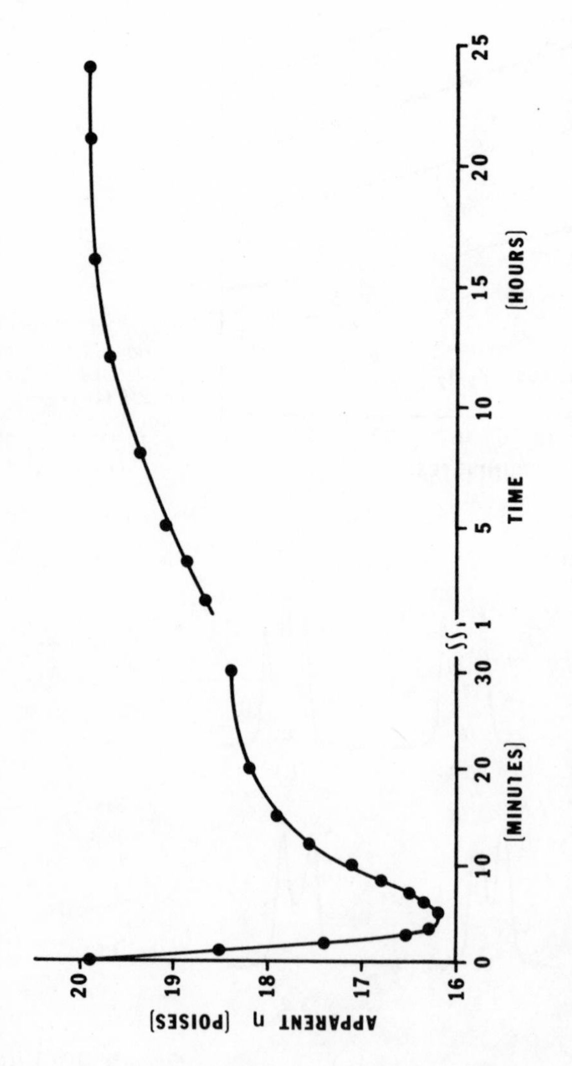

Figure 3. Apparent viscosity vs. time after a 90° reorientation of a 14.0% PBDG/CH_2Cl_2 sample in a 20-kG field at 25°C

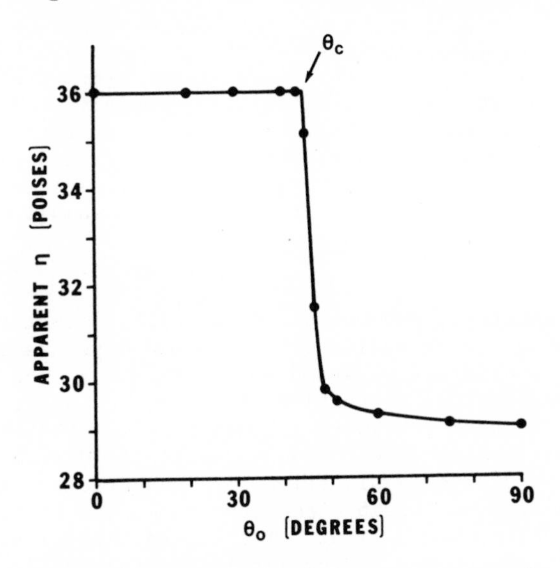

Figure 4. Apparent viscosity of a 24.5% sample after three consecutive reorientations of θ_o, each for 5 min, in a 17-kG field at 25°C

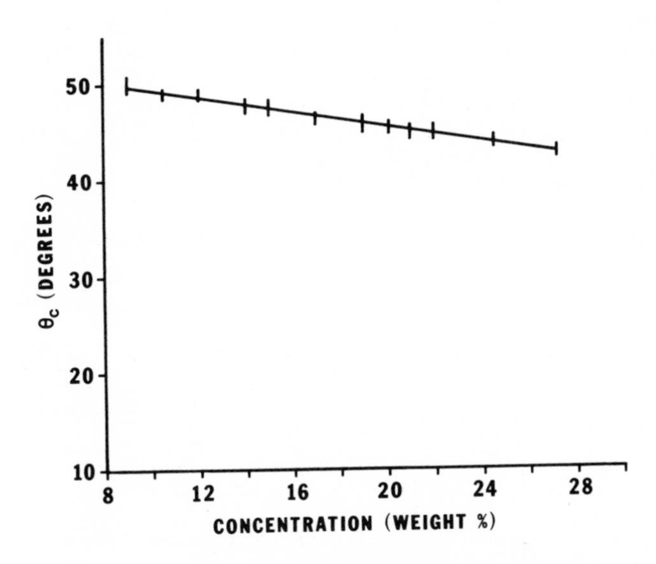

Figure 5. Concentration dependence of θ_c in PBDG/CH_2Cl_2 liquid crystals. Vertical bars represent error limits of θ_c.

linear and crossed circular polarizers, it is apparent that much of the color seen is due to the dispersion of form optical rotation, indicating the presence of twisted domains. Indeed, under certain conditions the disruptive effects of counterrotation can produce defects of a novel type, which are discussed elsewhere (16). When $\theta_o < \theta_c$, however, no disruptive effects are visible, even after several successive rotations of the sample.

Conclusions

NMR, viscometric, and optical data indicate that the mechanism of magnetic reorientation of PBG liquid crystals becomes complex above a critical rotation angle, $\theta_o = \theta_c$. The observed changes in the physical properties of the system under such conditions are all consistent with a model in which the PBG molecules are not perfectly aligned with the magnetic field at equilibrium. The consequence of such a situation in a magnetic reorientation experiment is to make possible the conditions for counterrotating regions to develop when $\theta_o > \theta_c$. The interaction of such domains can disrupt the structure of the liquid crystal and decrease the original degree of order of the polypeptide with commensurate changes in the NMR signal, apparent viscosity, and optical properties.

Literature Cited

1. Elliott, A., and Ambrose, E. J., Discuss. Faraday Soc. (1950) 9, 246.
2. Robinson, C., Trans. Faraday Soc. (1956) 52, 571.
3. Robinson, C., Ward, J. C., and Beevers, R. B., Discuss. Faraday Soc. (1958) 25, 29.
4. Robinson, C., Tetrahedron (1961) 13, 219.
5. Robinson, C., and Ward, J. C., Nature (1957) 180, 1183.
6. Sobajima, S., J. Phys. Soc. Japan (1967) 23, 1070.
7. Panar, M., and Phillips, W. D., J. Am. Chem. Soc. (1968) 90, 3880.
8. Samulski, E. T., and Tobolsky, A. V., Mol. Cryst. Liquid Cryst. (1969) 7, 433.
9. Fung, B. M., Gerace, M. J., and Gerace, L. S., J. Phys. Chem. (1970) 74, 83.
10. Filas, R. W., Hajdo, L. E., and Eringen, A. C., J. Chem. Phys. (1974) 61, 3037.
11. Faxen, H., Arkiv. f. Mat., Astron. och Fysik (1922) 17, 75.
12. Eringen, A. C., J. Math. Mech. (1966) 16, 1.
13. Eringen, A. C., Int. J. Eng. Sci. (1967) 5, 191.
14. Orwoll, R. D., and Vold, R. L., J. Am. Chem. Soc. (1971) 93, 5335.
15. Yokoyama, Y., Arai, M., and Nishioka, A., Polymer J. (1977) 9, 161.
16. Filas, R. W., J. Physique, in press (January, 1978).

RECEIVED December 8, 1977.

13

Mesomorphic Order in Block Copolymers from α-Amino Acids and Other Monomers and in Copolymers from α-Amino Acids

ANDRÉ DOUY and BERNARD GALLOT

Centre de Biophysique Moléculaire, C.N.R.S., 1 A, Avenue de la Recherche Scientifique, 45045 Orléans Cedex, France

For more than ten years we have been interested by the synthesis and the structure of block copolymers (1,2). In 1973, we have thought that copolymers with one or two polypeptide blocks would be of high interest both in a technological and in a biological point of view. Due to their ability to exhibit a large range of conformations, polypeptide chains should be able to confer to copolymers a large variety of new technological properties on one hand, and to mimic the behaviour of membranous proteins on the other hand (3). So we have synthesized and studied by X-ray diffraction, Electron Microscopy and Infrared Spectroscopy the following types of copolymers : copolymers with a carbohydrate and a peptide block, copolymers with a polyvinyl block and a polypeptide block, random and block copolymers of two different α amino acids.

In this paper, we give the principle of the synthesis of the three classes of copolymers and we sum up the principal characteristics of their mesomorphic structures.

Copolymers with a saccharide block and a peptide block

The first class of mesomorphic copolymers that we have synthesized and studied are copolymers with a hydrophilic saccharide block and a hydrophobic peptide block. In these copolymers, the saccharide block is a carbohydrate fraction of a glycoprotein. A glycoprotein can be described as a graft copolymer in which a small number of carbohydrate chains are grafted to a peptide skeleton.

Synthesis. We have used as glycoprotein Ovomucoid extracted from Hen egg white (4). The enzymatic degradation of Ovomucoid gives a mixture of carbohydrates. Their fractionation by column chromatography provides two glyco amino acids which are both terminated by an Asparagin residue (5). The α fraction (O_α) has a molecular weight of 1850 and contains 10 sugar residues ; the β fraction (O_β) has a molecular weight of 3200 and contains 16 residues (5). Using the α amine function of the terminal Asparagin residue

0-8412-0419-5/78/47-074-165$05.00/0

of the glyco-amino-acids O_α and O_β we have initiated the polymerization of the NCAs of benzyl-L-glutamate and cynnamyl-L-glutamate and prepared copolymers containing between 14 % and 86 % of polypeptide (6).

Structure. Block copolymers O_αEB and O_βEB, where the hydrophobic polypeptide block (EB) is a poly(benzyl-L-glutamate) block, exhibit mesophases in dimethyl sulfoxide for DMSO concentrations ranging from zero to a limit value which depends upon the composition of the copolymer and the nature of the carbohydrate block (2,6).

The study by low angle X-ray diffraction of these mesophases provides X-ray patterns exhibiting in their central region a set of sharp lines with Bragg spacings in the ratio 1,2,3,4,5.., characteristic of a layered structure. This lamellar structure results from the superposition of plane, parallel, equidistant sheets ; each sheet contains two layers : one of thickness d_A formed by the carbohydrate block, the other of thickness d_B formed by the polypeptide blocks ; there is a partition of the solvent between the two layers : 70 % of the solvent is localized in the carbohydrate layer (6). Furthermore, in the polypeptide layer, the peptide chains are in an α helix conformation as is demonstrated by Infrared spectroscopy and X-ray diffraction and are assembled on a bidimensional hexagonal array for copolymers with an O_β saccharide block as is revealed by X rays (6).

When the molecular weight of the polypeptide block increases, the total thickness of a sheet and the thickness d_B of the polypeptide layer both increase linearly while the thickness d_A of the carbohydrate layer remains nearly constant.

The effect of the solvent concentration on the geometrical parameters of the lamellar structure is similar for all the copolymers studied and is illustrated in Fig. 1 for the copolymer O_β EB.33 which contains 62 % of polypeptide. When the DMSO concentration increases : the total thickness d of a sheet increases, the thickness d_A of the carbohydrate layer also increases, while the thickness d_B of the polypeptide layer remains nearly constant.

Copolymers with a polyvinyl block and a polypeptide block

The second class of mesomorphic copolymers that we have synthesized and studied consists of copolymers with a polyvinyl block and a polypeptide block. In these copolymers the first block is a polystyrene or a polybutadiene block and the second block is a hydrophobic or a hydrophilic polypeptide block.

Synthesis. The polyvinyl block (polybutadiene or polystyrene) is synthesized by anionic polymerization, then the chemical modification of the living ends provides a polymer terminated by a primary amine function which is used to initiate the polymerization of the NCA of the desired α amino acid (7,8).

By this way we have prepared the following copolymers : polybutadiene-poly(benzyl-L-glutamate) (BG), polystyrene-poly(benzyl-L-glutamate) (SG), polybutadiene-poly(carbobenzoxy-L-lysine) (BCK), polystyrene-poly(carbobenzoxy-L-lysine) (SCK), polystyrene-poly(L-leucine) (SL), polybutadiene-poly(N^5-hydroxypropyl-L-glutamine) (BGH), polybutadiene-poly(L-lysine) (BK), polystyrene-poly (L-lysine) (SK) and polystyrene-poly(L-glutamic acid) (SE).

Structure. Copolymers BG, SG, SL, BCK and SCK exhibit liquid crystalline structures in the dry state and in concentrated solution in different solvents : dioxane, dichloro ethane, dichloro propene...

For all copolymers studied (copolymers containing between 18 % and 83 % of polypeptide) the liquid crystalline structures are always lamellar and are very similar to the structure of saccharide-peptide block copolymers. In the lamellar structure of copolymers with a polyvinyl block and a hydrophobic polypeptide block, each sheet of thickness d results from the superposition of two layers : one of thickness d_A formed by the polyvinyl chains in a more or less random coil conformation, the other of thickness d_B formed by the polypeptide chains, in an α helix conformation, arranged in an hexagonal array, and generally folded (2,7,8).

The lamellar character of the structure, the α helix conformation of the polypeptide chains and their hexagonal packing are demonstrated, as in the case of saccharide-peptide block copolymers, by X-ray diffraction, Electron microscopy (Fig. 2), Infrared spectroscopy and Circular Dichroism (7). The number of folds of the polypeptide chains is determined by comparison of the thickness d_B of the polypeptide layer with the average length L of the polypeptide chains calculated from their degree of polymerization and their α helix conformation (2,7). The number of folds of the polypeptide chains is governed by the molecular weight and the composition of the copolymer and by the nature of the blocks. The number of folds increases with the polypeptide content of the copolymer : for instance, for copolymers SG of different composition but with all a molecular weight of 25000 for the polystyrene block, the number of folds increases from 0 for a peptide content of 31 % to 5 for a peptide content of 71 %. Furthermore, the number of folds is higher if the polypeptide chain is linked to a polystyrene chain than to a polybutadiene one (9).

Fig. 3 illustrates the behaviour of the geometrical parameters of the lamellar structure versus solvent concentration for copolymers with a polyvinyl block and a hydrophobic polypeptide block. As in the case of saccharide-peptide copolymers, when the solvent concentration increases, the total thickness d of a sheet and the thickness d_A of the polyvinyl layer both increase, while the thickness d_B of the polypeptide layer remains nearly constant.

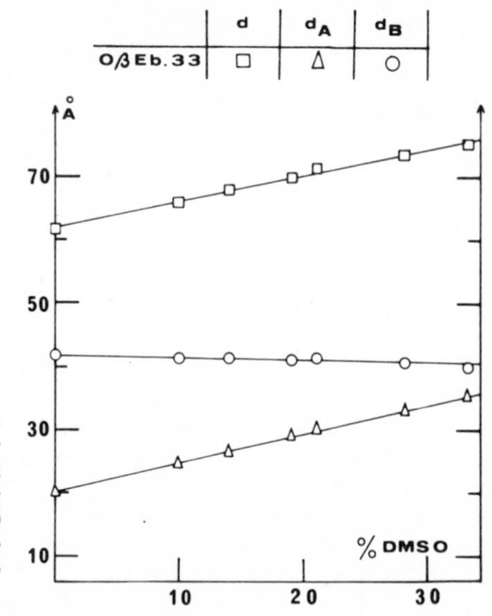

Figure 1. Variation with DMSO concentration of the parameters of the lamellar structure of the copolymer 0_βEB.33 with a saccharide block of the β type and a poly (benzyl-L-glutamate) content of 62%. (□) Total thickness d of a sheet; (△) thickness d_A of the carbohydrate layer; (○) thickness d_B of the peptide layer (6).

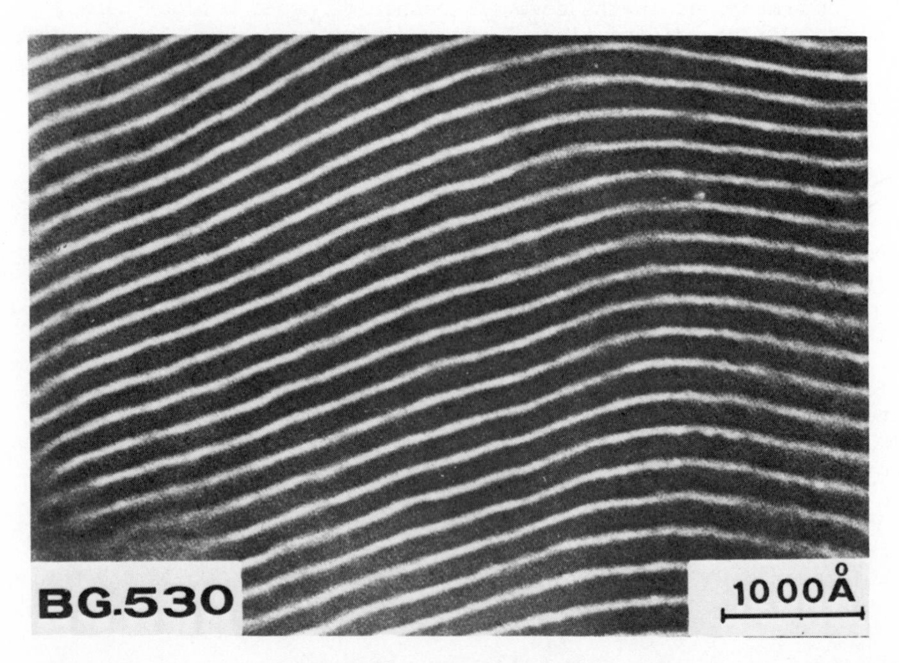

Figure 2. Example of electronic micrographs provided by copolymers polybutadiene–poly(benzyl-L-glutamate). White stripes, polypeptide layers; black stripes, polybutadiene layers stained with Osmium.

Random and block copolymers of two α amino acids

The third class of mesomorphic copolymers we have synthesized and studied are random and block copolymers of two different α amino acids.

Synthesis. Using hexamethyl amine as initiator, we have copolymerized, in DMF solution, the N-carboxy-anhydride (NCA) of benzyl-L-glutamate with the NCAs of the following amino acids : cinnamyl-L-glutamate, carbobenzoxy-L-lysine, L-leucine, D,L-phenyl alanin, L-phenyl alanin.

To obtain block copolymers, we copolymerize successively the NCAs of the two α amino acids. At the end of the polymerization of the first block we protect the NH_2 terminus by an O-nitrophenyl-sulphenyl group to allow its fractionation and its characterization. After elimination of the protecting group we polymerize the NCA of the second amino acid.

To obtain random copolymers, we polymerize simultaneously the NCAs of the two amino acids and we follow the copolymer composition during polymerization to be sure that it remains nearly constant and that there is no formation of blocks.

Structure. All the random and block copolymers we have synthesized exhibit liquid crystalline structures in dioxane.

We have not studied the cholesteric phases found between about 50 % and 80 % of solvent for some copolymers, because X-Ray diffraction patterns of this phase are very poor (one observes only a band at low angles) and do not allow a detailed analysis of the cholesteric structure.

On the contrary, for dioxane concentrations smaller than about 50 % to 60 %, low angle X-Ray diffraction patterns exhibit sets of sharp lines which allow the determination of the structure of the mesophases.

In these mesophases, the polypeptide chains are in an α helix conformation as it is demonstrated by Infrared spectroscopy (from the position of the characteristic bands Amide I and Amide II) and by X-Ray diffraction (from the value of the mass per length unit (10)).

In these mesophases, the packing of the helices, deduced from low angle X-Ray patterns depends upon the type of copolymer (random or block), the composition of the copolymer and the nature of the blocks.

1) Homopolymers of benzyl-L-glutamate. Homopolymers of benzyl-L-glutamate exhibit, as it is well known, an hexagonal structure formed by α helices packed in a bidimensional hexagonal array. The distance between the axis of two neighbouring helices increases with the dioxane concentration as it is shown on the Figure 4. The domain of stability of the hexagonal structure varies slightly with the molecular weight of the polymer.

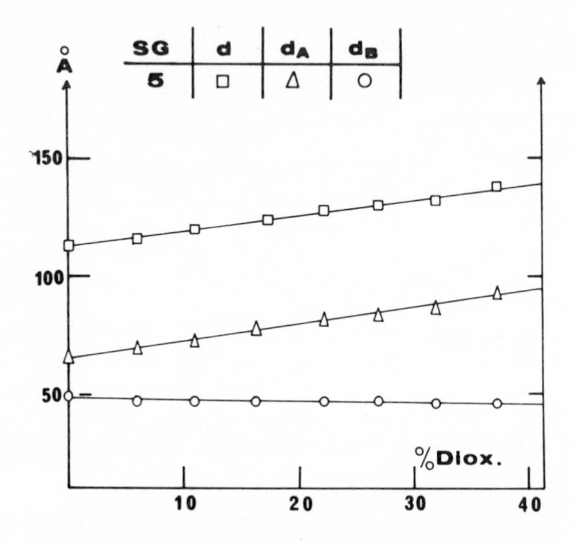

Figure 3. Variation with dioxane concentration of the parameters of the lamellar structure of the copolymer SG.5 with a polystyrene block of 25,000 and a poly(benzyl glutamate) content of 47%. (□) Total thickness d *of a sheet; (△) thickness* d_A *of the polystyrene layer; (○) thickness* d_B *of the polypeptide layer.*

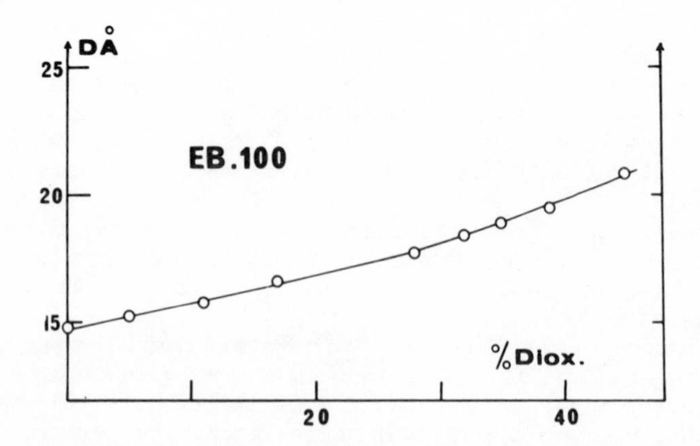

Figure 4. Variation with dioxane concentration of the parameter of the hexagonal lattice of a homopoly(benzyl-L-glutamate) with a number average degree of polymerization of 100

2) Homopolymers of carbobenzoxy-L-lysine. We have studied homopolymers of carbobenzoxy-L-lysine with an average degree of polymerization between 50 and 500. They exhibit in dioxane solution different structures as it has already been reported for other solvents (11)

For dioxane concentrations between 55 and 70 %, X-ray patterns are characteristic of a cholesteric structure.

For dioxane concentrations between 40 and 55 % and for dry copolymers X-ray patterns exhibit reflexions characteristic of an hexagonal structure (Bragg spacings in the ratio 1, $\sqrt{3}$, $\sqrt{4}$, $\sqrt{7}$, $\sqrt{9}$..).

For dioxane concentrations between 9 and 23 % X-ray patterns exhibit a set of reflexions which can be indexed in a quadratic lattice (Bragg spacings in the ratio 1, $\sqrt{2}$, $\sqrt{4}$, $\sqrt{5}$, $\sqrt{8}$..). At both ends of the domain of stability of the quadratic structure one observes a small domain of concentration where a demixion between quadratic and hexagonal structures exists.

For dioxane concentrations between 25 and 40 %, if the samples are prepared in standard conditions, X-ray patterns exhibit a diffuse band which does not allow the determination of the parameter of the bidimensional lattice ; nevertheless Infrared spectroscopy reveals the existence of helices.

In order to resolve the structure of the polymer in this domain of concentrations we have used the effect of temperatures on the stability of the different structures. We have prepared at 80°C mesomorphic gels containing 50 % of dioxane and exhibiting a well definite hexagonal structure, cooled the sample at - 30°C, evaporated the solvent at - 30°C until the desired concentration and annealed the sample at 20°C. By this way we have obtained a well organized hexagonal structure for dioxane concentration higher than 23 %. We have even been able to obtain a quadratic structure for the dry copolymer by evaporation at - 30°C of the solvent of a sample containing 15 % of dioxane.

The quadratic structure is stable in a small domain of concentration and temperature. Its highest stability is observed for dioxane concentrations of about 15 % : the quadratic structure is stable until temperatures higher than 90°C. For other dioxane concentrations the transition quadratic → hexagonal is observed at temperatures only slightly higher than room temperature.

On the contrary, the hexagonal structure is stable until temperatures higher than about 200°C where degradation of the polymer begins.

If the hexagonal structure is particularly stable it is not the only structure possible for polymers of carbobenzoxy-L-lysine. For certain concentrations and temperatures a quadratic structure appears and it probably corresponds to a special conformation of the lateral chains. The elevation of the temperature increases the thermic agitation and destroys the asymmetry of the lateral chains leading to on hexagonal structure characterized by a cylindrical symmetry of the lateral chains around the peptidic chains.

Fig. 5 illustrates the respective domains of stability of the quadratic structure (triangles) and of the hexagonal structure (circles) at 20°C. It also shows that the lattice parameter of the two structures increases with dioxane concentration.

3) Copolymers of carbobenzoxy-L-lysine and benzyl-L-glutamate. We have studied a lot of random and block copolymers of carbobenzoxylysine and benzyl glutamate of different compositions. They all exhibit an hexagonal structure whose the lattice parameter increases with solvent concentration.

The figures 6 and 7 show that the laws of variation of the distance between the axis of the neighbouring helices versus solvent concentration are the same for random and block copolymers of the same composition.

The fact that random and block copolymers of the same composition present the same hexagonal lattice with the same geometrical parameters means that the helices can slide along the direction of their axis and that there is no phase separation between the poly(carbobenzoxy-L-lysine) and the poly(γ-benzyl-L-glutamate) blocks.

On figure 8 we have plotted the variation of the parameter of the hexagonal lattice for homopoly(benzyl-L-glutamate) : curve D, for homopoly(carbobenzoxy-L-lysine): curve A and for random and block copolymers of different compositions : curve B (copolymers containing 33 % of carbobenzoxy-L-lysine) and curve C (copolymers containing 66 % of carbobenzoxy-L-lysine). One can see that the distance between the axis of the helices increases with the carbobenzoxy-lysine (CK) content of the copolymer.

4) Copolymers of L-leucine and benzyl-L-glutamate. For all random and block copolymers of L-leucine and benzyl-L-glutamate that we have synthesized and studied the polypeptide chains are in an α helix conformation.

Random copolymers of L-leucine and benzyl-L-glutamate exhibit an hexagonal structure as homopolymers of benzyl-L-glutamate do.

Block copolymers of L-leucine and benzyl-L-glutamate (LEB) containing less than 30 % of L-leucine exhibit an hexagonal structure.

Block copolymers L.EB containing between 40 and 60 % of L-leucine exhibit a centred rectangular structure. This structure is a deformation of an hexagonal structure (see Fig. 9) and each elementary lattice contains two chains : one of each type of block. This fact is in agreement with the domain of copolymer composition which gives this centred rectangular lattice.

The variation of the parameters of the centred rectangular array versus dioxane concentration is shown is figure 10. One can see that one side of the array increases much faster than the other.

Conclusion

Among the mesomorphic copolymers that we have synthesized and

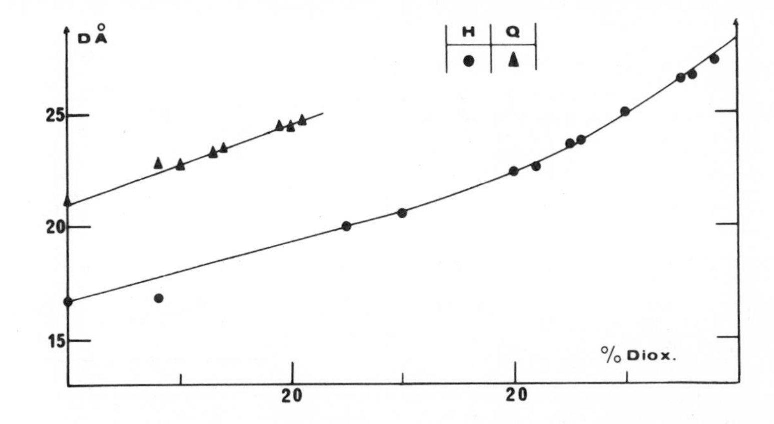

Figure 5. Variation with dioxane concentration of the parameter of the hexagonal lattice (●) and of the quadratic lattice (▲) of polymers of carbobenzoxy-L-lysine

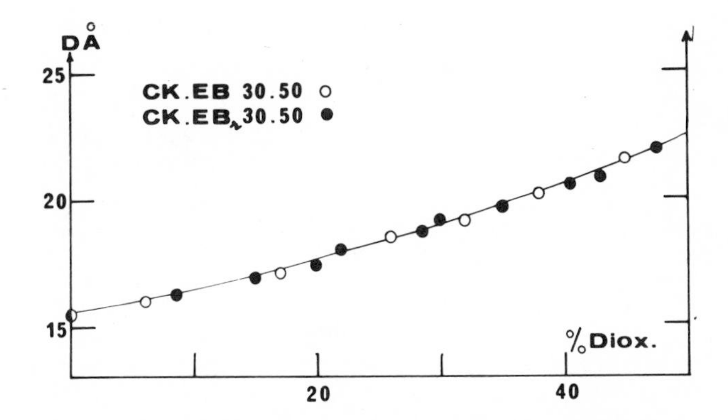

Figure 6. Variation with dioxane concentration of the parameter of the hexagonal lattice for random and block copolymers with number average degree of polymerization of 30 for carbobenzoxy lysine and 50 for benzyl glutamate. (○) Block copolymer CK.EB.30.50; (●) random copolymer CK.EB$_r$.30.50.

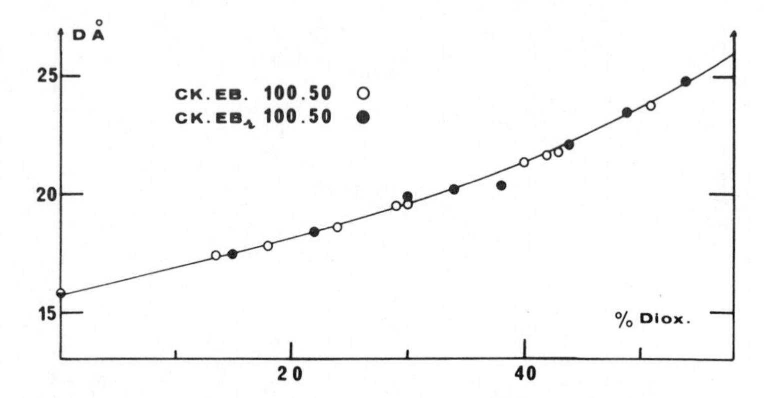

Figure 7. Variation with dioxane concentration of the parameter of the hexagonal lattice for random and block copolymers with number average degree of polymerization of 100 for carbobenzoxy lysine and 50 for benzyl glutamate. (○) Block copolymer CK.EB.100.50; (●) random copolymer CK.EB$_r$.100.50.

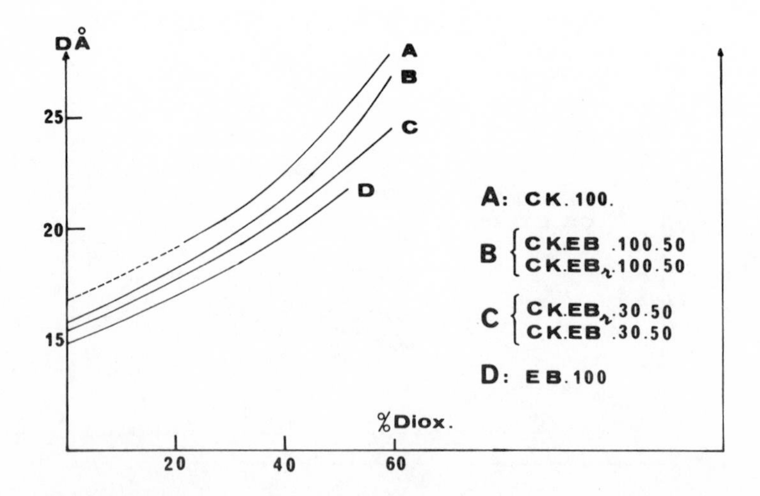

Figure 8. Variation with dioxane concentration of the parameter of the hexagonal lattice of homopolymers, block, and random copolymers of carbobenzoxy-lysine and benzyl glutamate. (D) homopolybenzyl glutamate EB.100; (C) block and random copolymers CK.EB.30.50 and CK.EB$_r$.30.50; (B) block and random copolymers: CK.EB.100.50 and CK.EB$_r$.100.50; (A) homocarbobenzoxy-lysine CK.100.

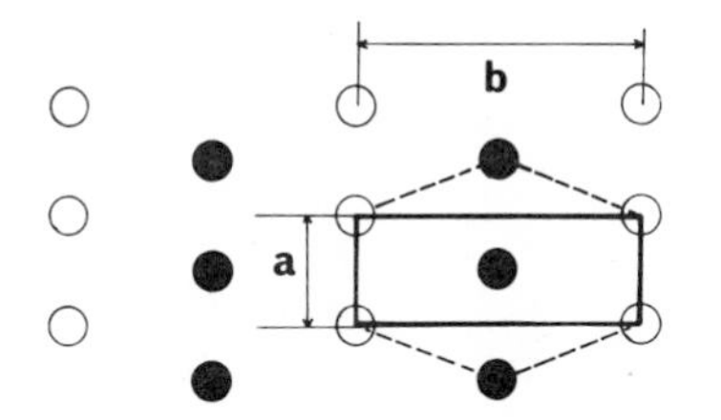

Figure 9. Schematic of the bidimensional rectangular array of poly(L-leucine)-poly(benzyl-L-glutamate) block copolymers (L.EB). (●) Axis of the helices of one type of chains; (○) those of the other type.

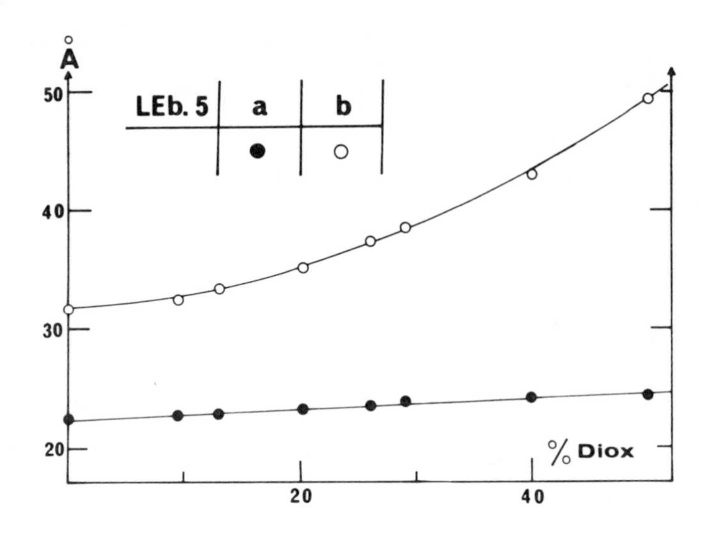

Figure 10. Variation with dioxane concentration of the parameters a and b of the centered rectangular array of the copolymer L.EB.5

studied, copolymers with a saccharide and a peptide block are of particular interest for biologists. They form simplified models of proteins, especially membrane proteins, according to their amphipatic character.

We are now preparing and studying membrane models formed by ternary systems : amphipatic block copolymer/lipids/water. From the interaction with our polymeric models of lectins (lectins are proteins or glycoproteins specific of different sugar residues) we hope to obtain informations about the respective parts played by the different carbohydrate chains and the polypeptide skeleton of glycoproteins and perhaps help to throw some light on problems as important as cell recognition and cell contact inhibition.

Abstract

The structure of random and block copolymers of two α amino acids, and of block copolymers with an amino acid block and a carbohydrate or a vinyl block have been studied by X-ray diffraction, electron microscopy, infrared spectroscopy and circular dichroism. Copolymers with a hydrophilic carbohydrate block and a hydrophobic polypeptide block exhibit in DMSO solution and in the dry state a lamellar structure ; this lamellar structure consists of plane, parallel, equidistant sheets ; each sheet results from the superposition of two layers : one formed by the carbohydrate chains in a disordered conformation, the other formed by the polypeptide chains in an helical conformation and assembled in a hexagonal or a centred rectangular lattice. Copolymers with a polyvinyl block (polystyrene or polydiene) and a hydrophobic polypeptide block exhibit a lamellar structure similar to that of copolymers with a carbohydrate and a peptide block, except that their polypeptide chains in an helical conformation are always hexagonally packed and are generally folded. For random and block copolymers of benzyl-L-glutamate and one of the following amino acids, ε-carbobenzoxy-L-lysine, L-leucine in the dry state and in dioxane concentrated solution the polypeptide chains are always in an helical conformation, but they are assembled on an hexagonal or a centred rectangular lattice depending upon the type of copolymer (random or block), the nature of the blocks and the copolymer composition.

Literature cited

1) Gallot Bernard, Pure Appl. Chem., (1974), 38, 1.
2) Gallot Bernard, "Liquid Crystalline Structure of Block Copolymers" in "Liquid Crystalline Order in Polymers", A. Blumstein Edit., Academic Press, New-York (1978).
3) Billot Jean-Pascal, Douy André, Gallot Bernard, Makromol. Chem., (1976), 177, 1889.
4) Lineaweaver H., Murray C.W., J. Biol. Chem., (1977) 171, 565.
5) Bayard Bernard, Thèse Doctorat, Lille, France (1974).

6) Douy André, Gallot Bernard, Makromol. Chem., (1977), 178, 1595.
7) Perly Bruno, Douy André, Gallot Bernard, Makromol. Chem., (1976), 177, 2569.
8) Billot Jean-Pascal, Douy André, Gallot Bernard, Makromol. Chem. (1977), 178, 1641.
9) Douy André, Gallot Bernard, paper in preparation.
10) Saludjian Pedro, Luzzati Vittorio, "Study of polypeptide-solvent systems" in "Poly α amino acids", G.D. Fasman Edit., Dekker, New-York (1967).
11) Saludjian Pedro, De Lozé Christiane, Luzzati Vittorio, C.R. Acad. Sci. Paris, 1963) C.256, 4297.

RECEIVED December 8, 1977.

14

Biomesogens and Their Models: Possibilities of Mesomorphic Order in Polynucleotide Analogues

S. HOFFMANN and W. WITKOWSKI

Department of Chemistry, Martin Luther University, DDR-40 Halle/S. and Central Institute of Microbiology and Experimental Therapy, Academy of Sciences of the GDR, DDR-69 Jena, East Germany

It seems that life - originated in the fluctuating interfaces of the primordial earth - escaped the threatening alternatives of a stiffening death within the sticky hierarchies of crystalline order and a senseless end of vanishing into the rather infinite fields of fluidity by occupying the small borderlines between the kingdoms of order and the surrounding chaos with amphiphilic species, fitted to those very housing conditions, enabled to use and amplify the advantages and to avoid and minimize the traps of this bipartite world, driven forward between the dialectics of excessive statics and dynamics into the developing, enlarging, optimizing spatial and temporal interconnective networks of complicated mesomorphic organizations. The ancient pool of nearly omnipotential macromolecules - built up from those amphiphilic moieties - developed into the specializations of informative, functional and compartimentive partners (Figure 1), conserving below the skin of their specific adjustments the continued primitive universality of the origin. Thus, while a first sudden glance might connect the structural features of nucleic acids with information, those of proteins with function and the remaining characteristics of membrane components with compartimentation, a nearer and more detailed intimacy with the three dominants of our today biopolymeric organizations seems to reveal much broader ranges of different abilities, merging the classical view of interacting structural individuals into the newer aspect of fluctuating mosaics of mutual domain cooperativities, where stereoelectronic patterns of interchanging parts of the competitors of the great old game anneal into space and time dependent networks of precisely tuned stereoelectronic fits of newly gained units of complex mesomorphic do-

0-8412-0419-5/78/47-074-178$12.50/0

main systems.

This view also seems to provide a better understanding for an as well basic as curious phenomenon, a special characteristic of biopolyelectrolytic systems: the time independent but directional hysteresis, that seems to have governed not only the first interchanges in the primordial omnipotential polyelectrolytic pool, but ruled, accelerated and directioned the driving forces of information processings on to the optimizations into the very regions of the central nervous systems of our species. The increasing efforts of the last years in elucidating those effects are dominated mainly by the basic investigations of Neumann and Katchalsky (1-4). Following the acid-base titration stimulus of Cox and coworkers (5) in the case of rRNA, they demonstrated the importance of generally long-lived metastabilities and particularly hysteresis phenomena in biopolyelectrolytic organizations for both regulation phenomena and memory imprints in bioorganisms. In their hands the 2poly(U)·poly(A) triplex advanced as a first star model not only for the investigation of hysteresis phenomena by spectrophotometric as well as potentiometric acid-base titrations, but also as a first demonstration object for inducing long-lived conformational changes in metastable biopolymeric organizations by electric field pulses, surprisingly in the order of the action potentials of the nervous membrane (3,4,6).

From this and further investigations a new intriguing dynamic picture of biosphere emerged, where biopolyelectrolytic systems, sensitized by their outstanding characteristics of exhibiting metastabilities and molecular hystereses to passive regulations from environmental electromagnetic influences and vice versa active mediations into self-governed areas, gain the possibilities of elementary cybernetic units with the special facilities of transducer functions in general and memory recordings and stimulations of biorhythms in particular (2,7,8). The list of examples and applications has been enlarged in the meantime not only in the field of polynucleotides (9-14), but also in the widespread areas of proteins (15-17) and membrane components (18-23), thus underlining the early prediction of metastabilities and molecular hysteresis phenomena as common features to all biopolyelectrolytic moieties. In these days they cumulated in the interesting and important results of a first demonstration as a working principle in a biological process. Schneider and coworkers (24,25) demonstrated by means of a special flow reactor the fascinating, but yet somehow expected dependence of the accuracy of enzymic reading a template from the very way of doing this, thus providing for the first time evidence for the capabilities of hysteresis in governing not only the space, but also the longly forgotten time dependent hierarchies of bioregulations. While these results seem to open a new field in studying chronobiological interrelationships on the level of interacting native systems, our own efforts in this context started and remained up to now far below those levels of high-ranking ambitions (7,14,26).

Stimulated by the early efforts of Langenbeck (27) to create and build up some sort of biomimetic chemistry at a time when these aspects had scarcely been noticed, we intended to start a program of modelling certain aspects of the hitherto yet inaccessible jungles of native biopolymeric interactions. In this context we tried to add further evidence for molecular hystereses of some polynucleotide triplexes (7,12-14), that might display some regulatory functions in biological viro-, cancero- and immunemodulations (28-32). Looking for the basal features of base-stack and strand-pattern interrelations and interaction possibilities, it seemed to us, that the widely enlarging chemistry of thermotropic mesogens (33) - especially in the case of using heterocyclic relatives (34,35) of nucleic acid bases as central parts - should provide rather good models, fitting some simpler aspects of thermotropic behaviour in the hydrophobic mesomorphic core region of polynucleotides. Yet the clearly distributed molecular mesogen characteristics of our synthetic liquid crystals and their theoretical extrapolations into the ordered interaction fields of large ensembles do not only seem to overemphasize rather secondary and logically delicate subcategorizations (lyotropy and thermotropy), but might moreover block their own scientific development into the cooperative domain systems of mesogen biopolymeric organizations. To bridge this gap between symmetry in vitro models and asymmetrically directed in vivo realities, it seemed tempting, not only to study polymeric mesogens - mainly built up from classical mesomorphic monomers - as it has been excellently demonstrated by the groups of Blumstein, Ringsdorf, Cser, Hardy, Amerik, Strzelecki, Tanaka a.o. (36-46), or to look for mesogen areas even in technical products (47), but moreover to use simplified potentially mesogenic biopolymeric systems - "mesogen" here in the sense of domain characteristics rather than in describing larger ensemble areas - for modelling parts of these complex natural interdependences. Thus within the framework of our interests in metastabilities and hysteresis phenomena in bioregulations, we tried to model nucleic acids from a greater distance and to check these simulations in (bio)physical and biological interaction studies. The soon emerging possibilities for practical applications then led to a more general conception of strangening nucleic acids on different levels of stereoelectronic fit, utilizing both native monotonic sequences and synthetical mono- and oligomeric base-pair analogues as well as polymeric strand analogues in competitive and allosteric interaction studies (Figure 2). Modelling and varying the mesogen domain compositions of native systems not only by small nonmesogenic antimetabolites but also by at least partially mesogenic antitemplates (Figure 3), we hoped to adjust our effectors to their wanted levels of efficiency rather than to misuse them for a pure jamming of metabolic pathways.

Materials and Methods

Polynucleotides The polynucleotides were commercial products, obtained from Serva-Biochemica and Reanal. They have been used without further purification. $(A)_n$ (3.4/>8 S); $(U)_n$ (>10 S); $(G)_n$ (~10 S); $(C)_n$ (8-13 S); $(I)_n$ (10-15 S). Concentrations of polynucleotides in 0.01 M phosphate/ 0.1 M NaCl (PBS/ pH 7.2) were determined spectrophotometrically by using the following extinction coefficients: $(A)_n$: ε_{max}=10.000; $(U)_n$: ε_{max}=9.430; $(G)_n$: ε_{max}=9.600; $(C)_n$: ε_{max}=6.150; $(I)_n$: ε_{max}=10.400.

CT-DNA Calf-thymus DNA was a kind gift of Dr.G.Luck (Central Institute of Microbiology and Experimental Therapy, Academy of Sciences of the GDR) with a content of 0.28 % of protein, stabilized with chloroform. The stock solution was diluted to an OD_{260} of 0.625 by means of a 0.001 M NaCl solution.

Monomeric Thermotropics Long-chained 2-pyridone-5-carboxylic esters were prepared from 2-pyridone-5-carboxylic acid and the corresponding half ethers of hydroquinone by esterification with N,N'-carbonyldiimidazole according to the Staab-procedure (48). Detailed descriptions will be given elsewhere (49).

Long-chained esters of enantiomeric estradiols and long-chained ester-aniles of enantiomeric estrones were prepared by methods previously described (50,51,53). The enantiomeric estrones and estradiols

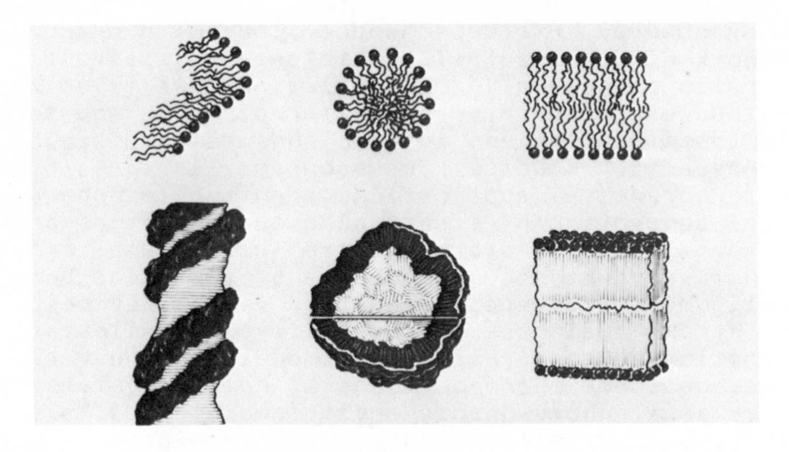

Figure 1. Amphiphilic pattern development in nucleic acids, proteins, and membranes

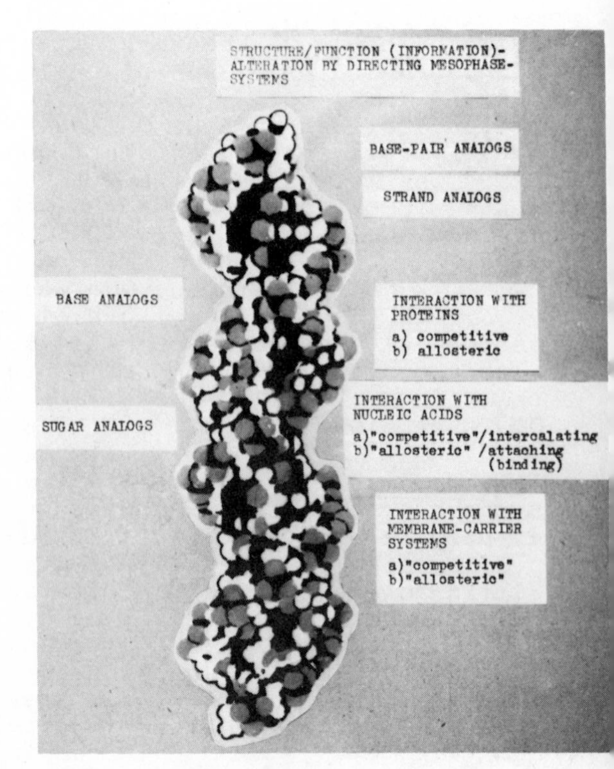

Figure 2. Nucleic acids "strangening" program

were purchased from VEB Jenapharm. Further details will be reported elsewhere (54,55).

Base-Pair Analogues 2.7-bis-[(dialkylamino- and {oxa}oligoalkylenimino)-acetylamino]-fluoren-9-ones ("fluoramides" 1-5) were prepared via the corresponding 2.7-bis-[(chloroacetyl)-amino]-fluoren-9-one by a procedure described recently (56); see also (57).

Vinylmonomers for Homo- and Copolymerization Reactions 1-Vinyluracil, 9-vinylhypoxanthine, and 9-vinyladenine were prepared by previously described methods (58-62), 1-vinylcytosine by the ammonolysis of 1-vinyl-4-methylthiopyrimidine-2-one (62,63).

Polynucleotide Strand-Analogues The following polymers (Tables I and II) were prepared by AIBN- or γ-ray induced (Laboratory Radiation Unit RCH-γ-30 as a ^{60}Co-source) polymerization and copolymerization reactions, respectively, followed in case of (vA, [vOH]$_{0.7}$)$_n$ by a polymeranalogous reaction. The starting amounts of N-vinylnucleobases and their comonomers were adjusted to yield a nearly 1:1-ratio in the copolymers.

Sedimentation Experiments Sedimentation measurements were carried out using a Spinco Model E ultracentrifuge, equipped with an ultraviolet absorption optic, a monochromator and a photoelectrical scanner, in a way previously described (66).

Base-Pairing Experiments Methods employed for base-pair complexation have been described previously (62,64,73,74). Ultraviolet mixing and thermal profile experiments were performed according to common procedures (73,74). Spectra for mixing curves were measured with an Unicam SP 1800 spectrophotometer (62,64,75). Melting curves were run on the same spectrometer, equipped with a SP 871 temperature program controler (62,64,75). ORD spectra were made on a Jasco ORD/UV-5 spectrometer, and the circular dichroic spectra were recorded on a Cary Model 60 spectrometer, equipped with a Model 6001 CD accessory (64,75,76).

Electron-Microscopic Studies Micrographs were taken at 36.000 · magnifications in an Elmiskop IA according to procedures previously given (65).

Hysteresis Experiments Molecular hysteresis measurements by cyclic acid-base titrations of the

Table I Homopolymers

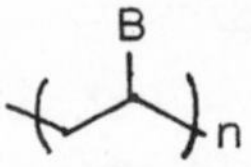

Polymer	Induction	B	Lit.
$(vA)_n$	AIBN	Ade	(58-61,67-70)
$(vU)_n$		Ura	(58-61,67-70)

Table II Copolymers

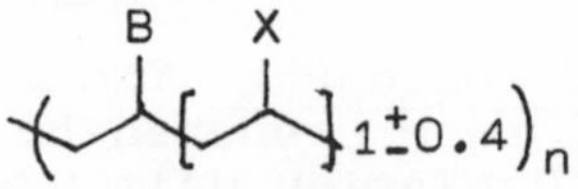

Copolymer	Induction	B	X	Lit.
$(vA,[vCOONa]_{0.8})_n$	AIBN	Ade	COONa	(66,68)
$(vA,[vCOONa]_{1.4})'_n$	γ	Ade	COONa	(66,68)
$(vA,[vSO_3Na]_{0.7})_n$	AIBN	Ade	SO_3Na	(75)
$(vA,[vOH]_{0.7})_n$	AIBN	Ade	OH	(65,66)
$(vA,[vCONH_2]_{1.1})_n$	AIBN	Ade	$CONH_2$	(75)
$(vA,[vPn]_{0.8})_n$	AIBN	Ade	2-pyrrolidone-1-yl	(75)
$(vA,[vP]_{0.9})_n$	AIBN	Ade	4-pyridyl	(75)
$(vA,[vI]_{1.1})_n$	AIBN	Ade	imidazol-1-yl	(75)
$(vA,[vm^3IJ]_{0.6})_n$	AIBN	Ade	3-methyl-[(imidazol-1-yl)-ium]-iodide	(75)
$(vU,[vCOONa]_{1.4})'_n$	γ	Ura	COONa	(67,68)
$(vC,[vCOONa]_{1.1})'_n$	γ	Cyt	COONa	(69,75)
$(vC,[vCOONa]_2)'_n$	γ	Cyt	COONa	(62,69)
$(vH,[vCOONa)'_n$	γ	Hyp	COONa	(62,75)
$(vH,[vCOONa]_{1.1})_n$	AIBN	Hyp	COONa	(69)

triplexes $(I)_n \cdot (A)_n \cdot (I)_n$ and $(U)_n \cdot (A)_n \cdot (I)_n$ were carried out as previously described (12-14,72), using the techniques of Neumann and Katchalsky (1,2,4). Titrations were performed in a modified USP-2 spectrometer of the Academy of Sciences of the GDR, with a pH ranging from 7.1 to 3.25.

DNA-Interaction Measurements DNA interactions with fluoramides were studied by DNA-titration with the intercalator. Melting profils were obtained on Unicam SP 1800 spectrophotometer and CD spectra on a Cary Model 60 as described above under "base-pairings" (77).

Biological Tests The polymers were tested as single strands in eukaryotic and viral polymerase systems. In base-paired complexes with their mate polynucleotides they were tested with regard to antiviral activities in establishing antiviral states, inducing and priming interferon.

CT-polymerase B (DNA-dependent RNA-polymerase) was obtained from calf-thymus tissues and assayed as reported earlier (71). Activity measurements were performed in a toluene system, using a Liquid Scintillation Counter Mark I (Nuclear Chicago).

AMV-Revertase (RNA-dependent DNA-polymerase) was assayed in an endogenous AMV-system as previously described (70). Avian Myeloblastosis Virus was a kind gift of Dr. J. Řiman (Institute of Organic Chemistry and Biochemistry, ČSAV, Praha, Czechoslovakia). Measurements were performed with a LKB-Wallace Liquid Scintillation Counter 81.000.

For estimations of antiviral activities (78) in general and interferon induction (62) in particular, an in vitro mouse L-cell system was used. VSV - (Vesicular Stomatitis Virus - strain Indiana) - was taken as a challenge virus. Yields of infectious virus were titrated by plaque assay on chicken embryofibroblasts.

For interferon priming experiments the effects of the semiplastics on NDV - (Newcastle Disease Virus - strain Hertfordshire)-mediated interferon induction in mouse L-cells was followed by the same common procedures previously described (78).

First orientating estimations of in vivo antiviral activities of fluoramides were performed in mice according to common procedures. Fuller details will be given elsewhere (79).

Molecular Hystereses in Polynucleotide Triplexes

When the beautiful Watson-Crick duplex emerged from a nameless mass of more or less complicated mono-, oligo- and polymeric structures, some longly missed esthetical features had been given back to organic chemistry, and this in both static and dynamic aspects (80). Figures 4, 5 and 6 contrast the less clearly arranged peptide backbones of cytochrome c (81), for instance, with the clearly distributed base-stack/strand-pattern arrangements of B-DNA and A-DNA or 11-RNA (28), respectively. While just the reduction of strandedness to the biologically promising duality granted the birth of this structure, nearly at the same time suggestions started or continued for multiple-strandedness in the biological materials (Figure 7). Four-strands structures, originally a theoretically based proposal of Gavin (90), have now become attractive as possible storage forms of DNA in the nucleosomes of chromatin (91). Yet considerably more interest has been spent in the meantime to triplexes, that had in former time been believed to represent the common strand design of DNA at all. Figure 8 reveals the appearance of the $(U)_n \cdot (A)_n \cdot (U)_n$ triplex according to the results of Arnott and coworkers (28). The originally rather theoretical aspects in dealing with those structure found a much more reliable and at the same time exciting basis, when poly(A) (7,92) opened the still continuing area of monotonic sequences in DNA and RNA interplays. Conserved even as it seems in some cases during evolution, these monotonics seem to play an important but up to now not yet elucidated role in bioregulations. The triplexes, which might arrange around those tracts in polynucleotides, remaining nevertheless structures still in want for definite biological functions, and in the meantime starting points or goals of a considerable number of different suggestions, ranging from a possible importance of those structure in early binary purine codes on the primordial earth (93), to the involvement in quite different regulations between replication, (reverse) transcription and translation up to the speculations for actions as cybernetic units (1-4,7-14,28) of a hysteretic regulation code acting by phase directing and manipulating strategies in bioregulations (7). The field of triplex structures - Figure 9 shows common pairing schemes - and their hysteretic behaviour, which had originally been the reservate of the working groups of Neumann and Katchalsky (1-4) as well as Guschlbauer (9-10), became a playground of endeavours

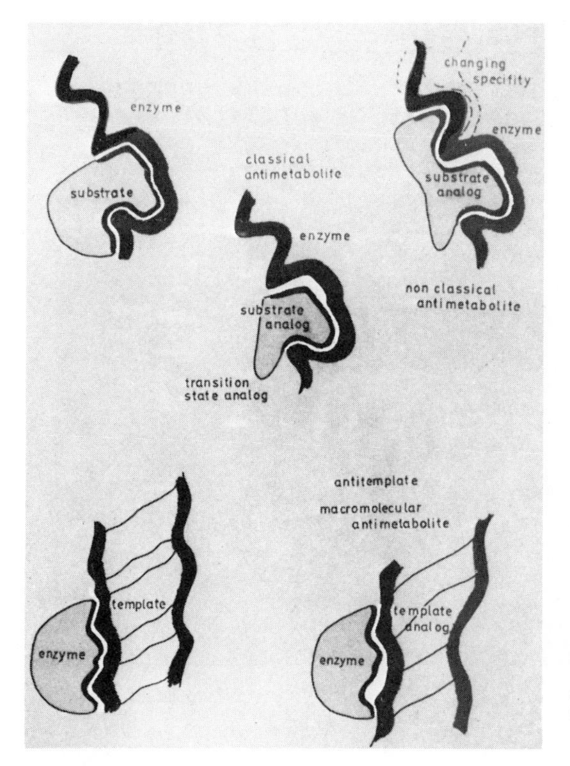

Figure 3. Survey of complementary matrix fits in nucleoproteinic systems

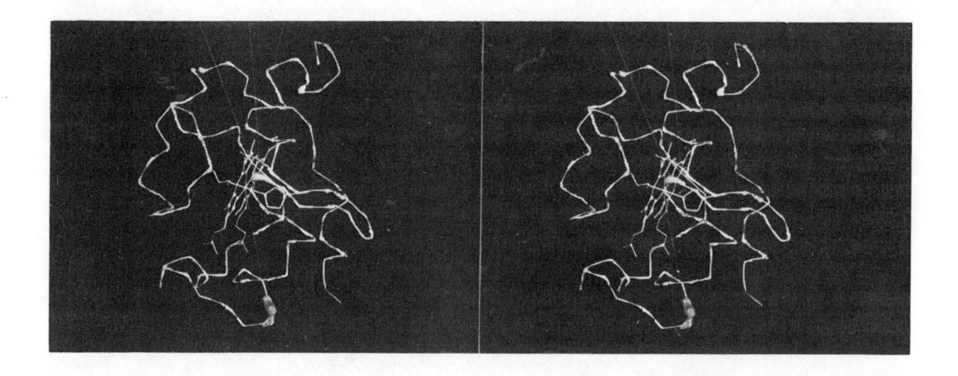

Journal of Biological Chemistry

Figure 4. αC-peptide backbone of cytochrome c (81)

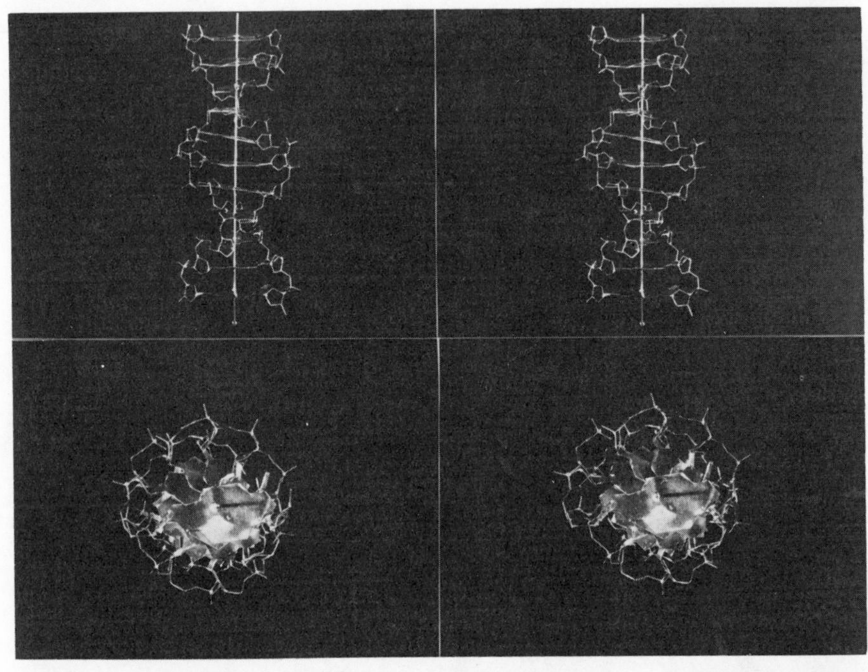

Nucleic Acids Research

Figure 5. (above) *B-DNA* (28)

Nucleic Acids Research

Figure 6. (below) *11-RNA* (28)

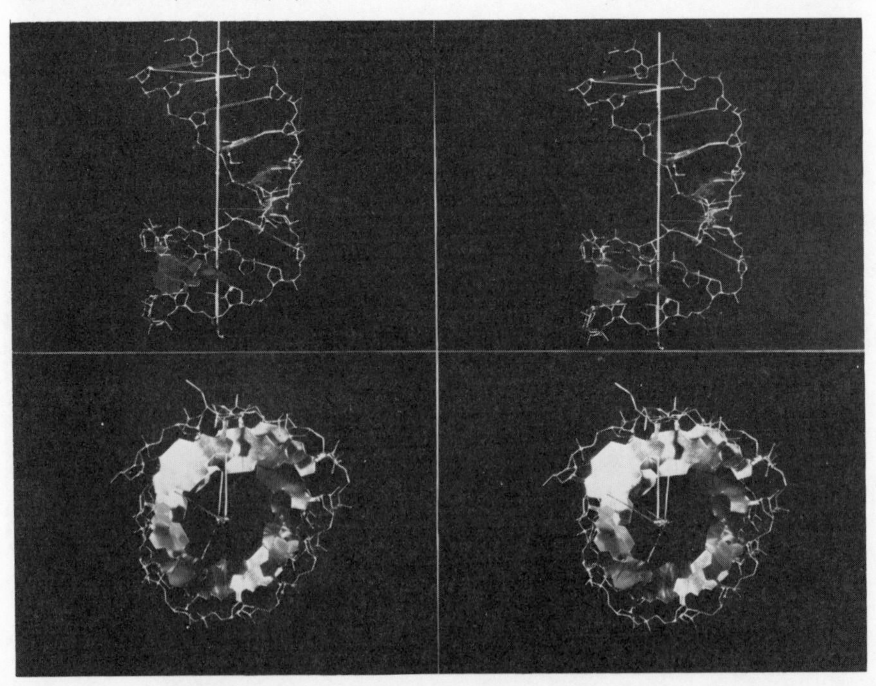

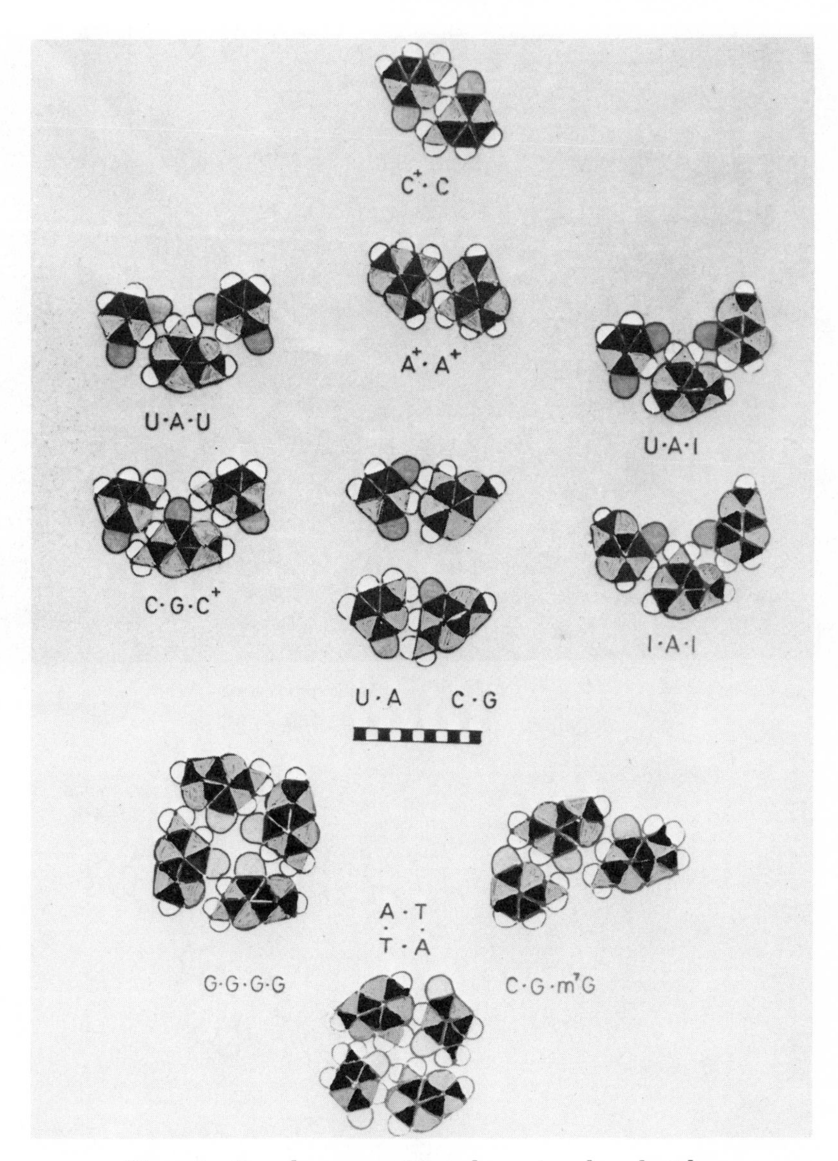

Figure 7. Base–base recognition schemes in polynucleotides

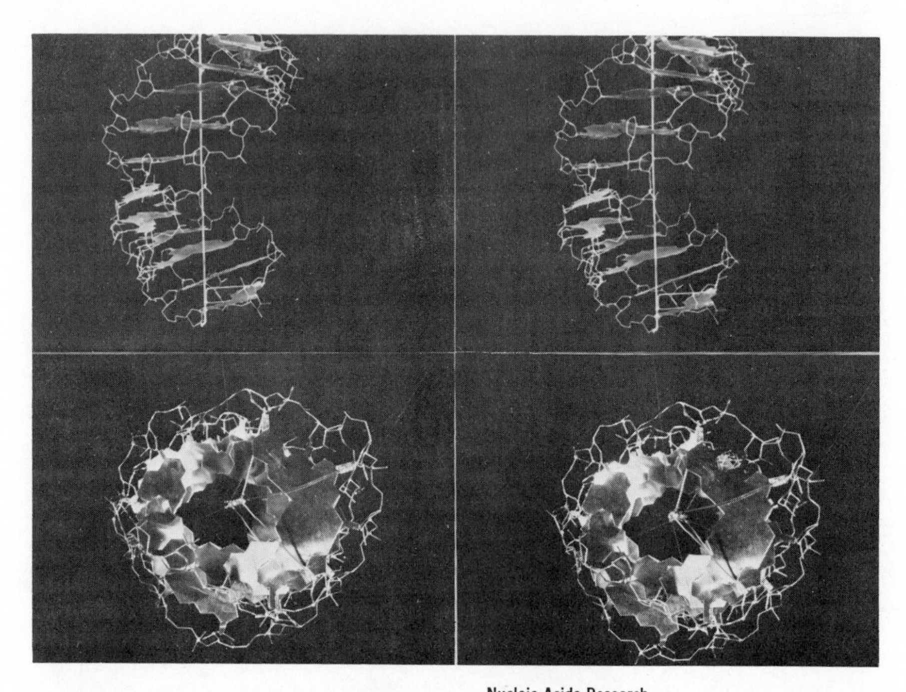

Figure 8. $(U)_n \cdot (A)_n \cdot (U)_n$*-triplex (28)*

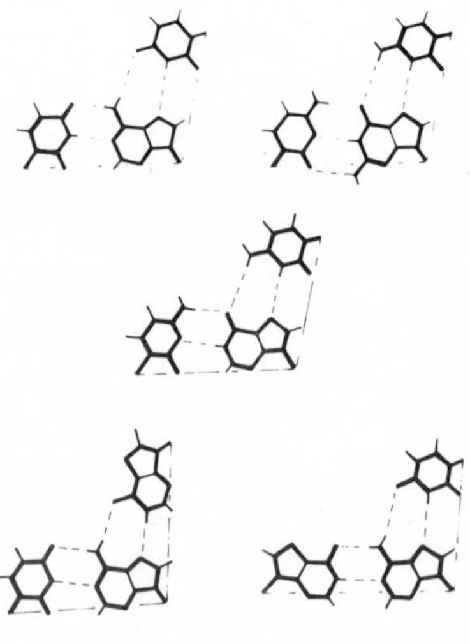

Figure 9. Triple-stranded polynucleotide pattern (28). (from top to bottom and left to right) *U·A·U, C·G·C⁺, C·I·C⁺, U·A·I, I·A·U, I·A·I; U·A·I/I·A·U-pattern in an as yet hypothetical arrangement* (94, 95).

from quite different points of view (7,28-32). We had been interested in those stereoelectronic pattern not only generally because of their possible involvements in bioregulations, but moreover particularly in Watson-Crick and Hoogsteen recognition ambitions of poly-(A)-monotonics with regard to poly(I)-tracts, due to the very effects of the latter in viro-, cancero- and immunemodulations (7,29-32,62).

After looking for a molecular hysteresis in the all-purine $(I)_n \cdot (A)_n \cdot (I)_n$ system (13), we spent interest to the hypothetical $(U)_n \cdot (A)_n \cdot (I)_n$ triplex (7) elucidated together with DeClercq and coworkers its existence (14,94,95) - Figure 10 shows melting profiles and ORD characteristics - and demonstrated also in this case of a mixed one-pyrimidine/two-purines-triplex a molecular hysteresis on cyclic acid-base titrations. Figure 11 and 12 summarizes our experiments in comparison with the pioneering study of Neumann and Katchalsky in case of $(U)_n \cdot (A)_n \cdot (U)_n$. For our two unusual triplex structures the hysteretic "lagging behind" of conformational orientations in their attempts to follow the pH-variations in the environment seems to be caused by similar transcrystallisation between the triplex and the $(A^+)_n \cdot (A^+)_n$ duplex, as proposed by Neumann and Katchalsky for the $(U)_n \cdot (A)_n \cdot (U)_n$ hysteresis, except that in our cases the process should be complicated in a still unknown way by the tendency to self-association of poly(I) (1,2,82,83). While from here there will arise the necessity to study those interrelationships much more thoroughly and on a larger scale, especially involving interactions between nucleic acids, proteins and membranes as well as their sensitivity to effector moieties and other environmental influences, we tried far below this level to introduce into the picture of bioeffector efficiencies (Figure 3) the above mentioned simulation concepts for providing antimetabolites and antitemplates derived from modelling conceptions in the field of nucleic acids and adjusted for mesomophic interactions with biopolymeric systems.

Modelling Nucleic Acids by Simple Thermotropics

The nicely balanced interplays of order-disorder features represented in the architecture of nucleic acids (Figure 13) (86,88,96), the complex intergame of hydrophobic and hydrophilic aspects, of weak and strong interactions - ranging from the hydrophobic long-core regions to the bizarre strand compositions with their different fits for handling nucleobase ac-

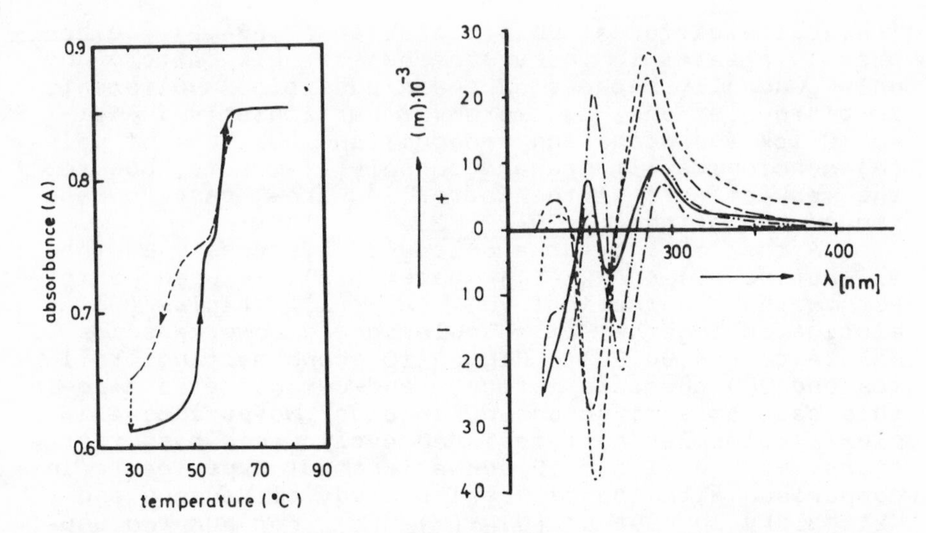

Figure 10. Thermal hysteresis in the melting profile and ORD-characteristics of $(U)_n \cdot (A)_n \cdot (I)_n$, the latter in comparison with the arrangement alternatives of the constituents. ORD-spectra: (——) $(U)_n \cdot (A)_n \cdot (I)_n$; (– – –) $(U)_n \cdot (A)_n \cdot (U)_n$; (·—·) $(I)_n \cdot (A)_n \cdot (I)_n$; (- - -) $(U)_n \cdot (A)_n$; (—··—) $(I)_n + (A)_n$.

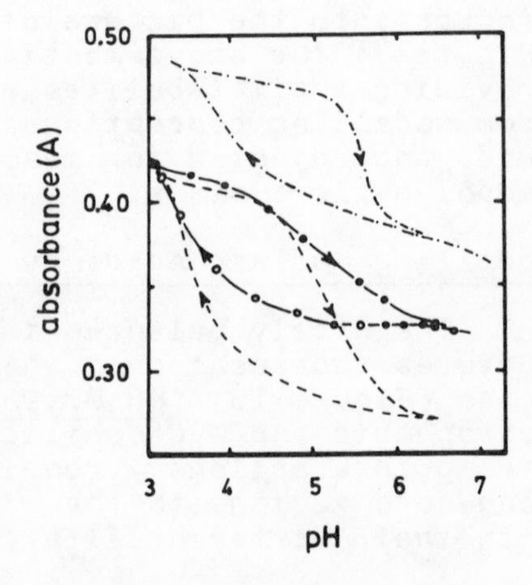

Figure 11. Molecular hysteresis in polynucleotide–triplex organizations. Cyclic spectrophotometric acid–base titrations. Absorbance (A) as a function of pH; c = 1.67 10^{-5} M triplex. (∘←∘—∘) Acid titration curve; (•→•—•) subsequent base-back titration curve; (– – –) $(U)_n \cdot (A)_n \cdot (U)_n$ after (1), A (260 nm), 0.1M NaCl, 0.005M PB; (——) $(U)_n \cdot (A)_n \cdot (I)_n$ after (14), A (250 nm), 0.15M NaCl, 0.01M PB; (—·—) $(I)_n \cdot (A)_n \cdot (I)_n$ after (12, 13), A (252 nm), 0.15M NaCl, 0.01M PB.

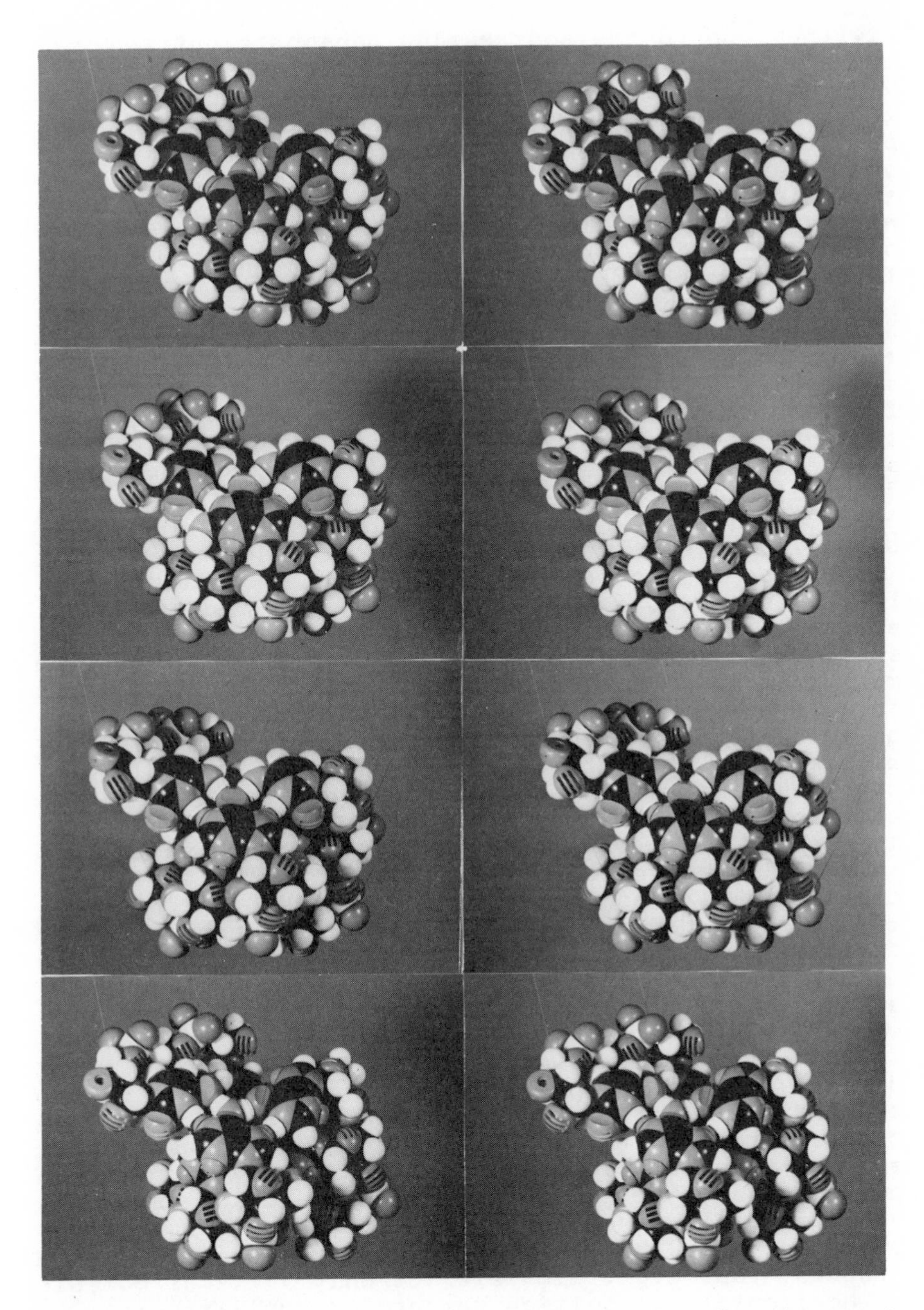

Figure 12. CPK-illustration of triplet parts of polynucleotide triplexes (28, 94, 95). (from top to bottom) $(U)_n \cdot (A)_n \cdot (U)_n$; $(C)_n \cdot (G)_n \cdot (C^+)_n$; $(C)_n \cdot (I)_n \cdot (C^+)_n$; $(U)_n \cdot (A)_n \cdot (I)_n$.

tivities, up to the contrasting phosphate pattern, the adjusted water shell and counterion clouds - express a vast amount of thermotropic and lyotropic characteristics and aspects, that might be modelled at least in parts by the much simpler stereoelectronic patterns of our today lyo- and thermotropic moieties. Added to this, the interaction facilities of nucleic acid bases with a lot of effector molecules - especially basepair analogues (97,98) - might offer a further field for approaching those mechanisms by simple thermotropic pattern, Figure 14 giving an of course, incomprehensive survey of common central parts of those structures. Thus, while a still enlarging number of substances is indicative of the lasting fruitful stimulation of the classical Kast structural proposal for thermotropic mesogens (33), the end of its dominance seems to come from some more general new trends in the field of mono-, oligo- and polymeric mesomorphs (1-4, 6-25,36-47,49-56). After the curiosities of the very beginnings, the widespread practical and theoretical fields opening to research within the last decades, there might be offered a new aspect for mesogen research in modelling complex bioregulations with a still promising feed back to their theoretical or even more practical developments. Hydrogen bonding systems, for instance, widely used in natural mesogens for quite different purposes, are still rather carefully avoided in synthetic mesogen chemistry and display a charming contrast between the differences of in vitro ambitions and in vivo realities. While in vitro only intramolecularly arranged hydrogen-bonding systems successfully occupied some reservates in the areas of classical Kast topological expectations, intermolecular hydrogen-bonding systems, with the exception of the indeed thoroughly investigated carboxylic acid dimers, remained the strange exceptions (33). Schubert and coworkers, interested in heterocyclic moieties as central parts for thermotropics on their ways of syntheses of low-melting nematics (34,35) - Figure 15 displays mainly a representation of their efforts - touched the field of those compounds, but without special interest in them. On the other hand we had in earlier times attached 2-pyridone-5-carboxylic acid to polyvinyl alcohol (99), thus building up some sort of very distantly modelled and abstracted polynucleotide analogue. From there we became interested in thermotropics, that might simulate not only vertical but also horizontal recognition pattern of nucleic acids. The here presented long-chained esters of the above mentioned 2-pyridone-5-carboxylic acid, given in a

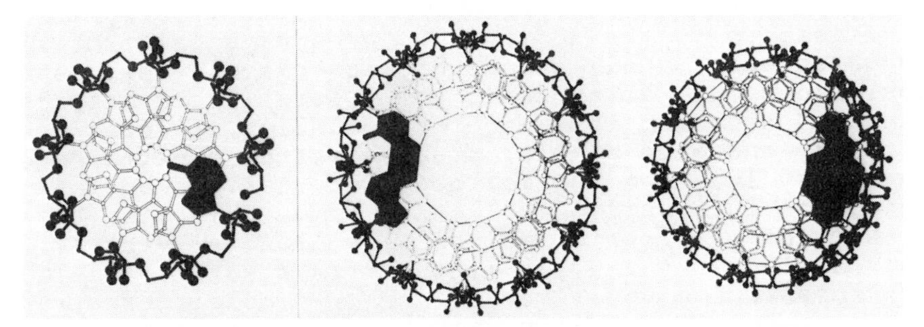

Figure 13. Amphiphilic mesomorphic pattern of single $(A)_n$ (88) *and double-stranded* $(U)_n \cdot (A)_n$ *and* $(G)_n \cdot (C)_n$ (96) *polynucleotides* (left to right)

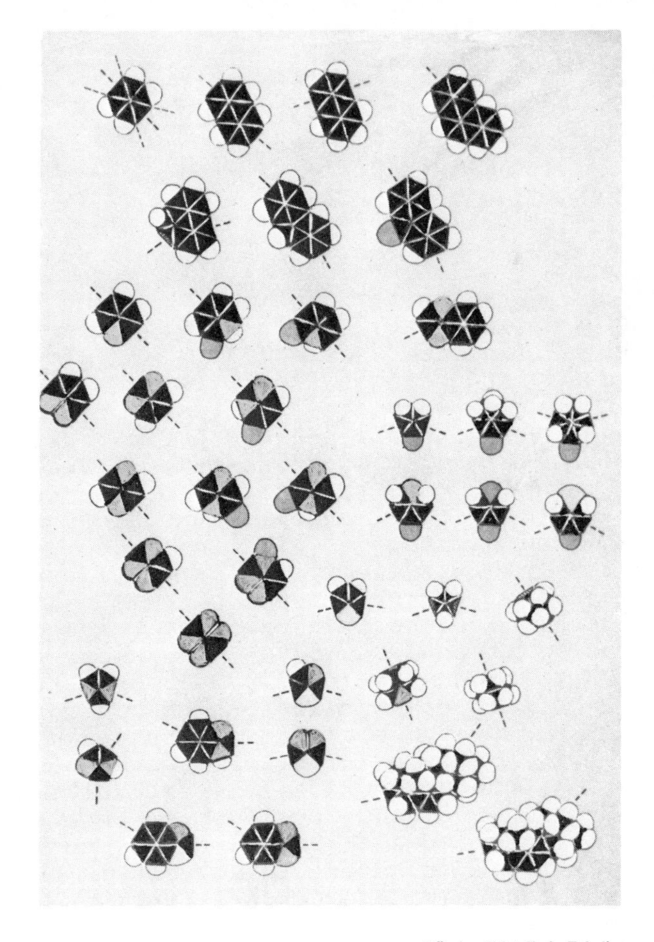

Flüssige Kristalle in Tabellen

Figure 14. Common central parts of classical thermotropics (33)

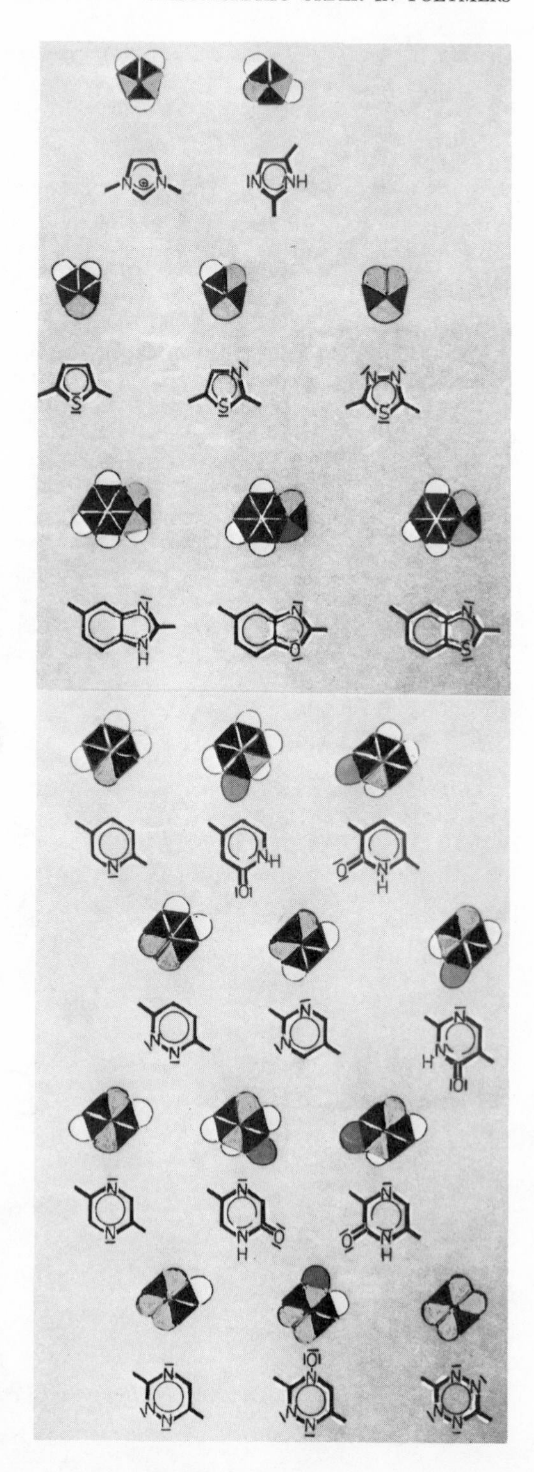

Figure 15. Heterocyclic central parts of classical thermotropics (33, 34, 35)

still speculative dimer arrangement with their smectic anisotropy areas in Figure 16 (49), might reveal at this moment a both doubting and intriguing glance into a still rather unknown field of biomesogen simulation possibilities.

Hereupon, adding to the more static view of those system a more dynamic aspect, it was only a short step to look also for thermotropic behaviour of components, that display interaction activities with nucleic acid bases, so for instance among other base-pair analogues: the stereoidal hormones (33,50-55,100-104). While their special mechanisms, sites and ways of efficiency - engaged with different receptors, their entrance and passage through membranes and compartimentations, reaching the final goal within chromatin - even at present remains rather unclear (101), those structures have been used in model experiments as "reporter molecules" for studying static and dynamic features of nucleic acids by employing their stacking or at least partially intercalating possibilities into nucleobase stacks (105,106). On the other hand in thermotropic mesogen chemistry a vast amount of cholesterol derivatives building up a kingdom of cholesterics is met only by a few examples of other steroidal groups as central part arrangements (Figure 17) (33,103). In the meantime we could add evidence for the thermotropic characteristics of the - though as it seems stereoelectronically predestinated, yet forgotten - estrogens (50-55), that will get cholesteric mesogen behaviour by simple lengthening their slightly distorted araliphatic central part geometry by commonly used terminal design, and extend those investigations to the enantiomeric partners and racemic mixtures (Figure 18 and 19) (54,55).

These two limited examples might continue the hopes, that also in this way some fruitful interchange of static as well as dynamic in vivo and in vitro aspects of mesomorphic pattern might stimulate and enlarge a better vice versa understanding of both artificial and native mesogen pattern in the biosphere.

"Strangening" Nucleic Acids by "Infective" Compounds

"Wir haben nun gefunden, daß gereinigte, kristallisierte Abietinsäure in hervorragendem Maße die Fähigkeit hat, optisch inaktive flüssige Kristalle in zirkular polarisierende umzuwandeln. Minimale Mengen der Säure verursachen eine optisch zirkulare Infektion, durch welche die Gesamtmasse der an sich inaktiven flüssigen Kristallschicht stark zirkular polari-

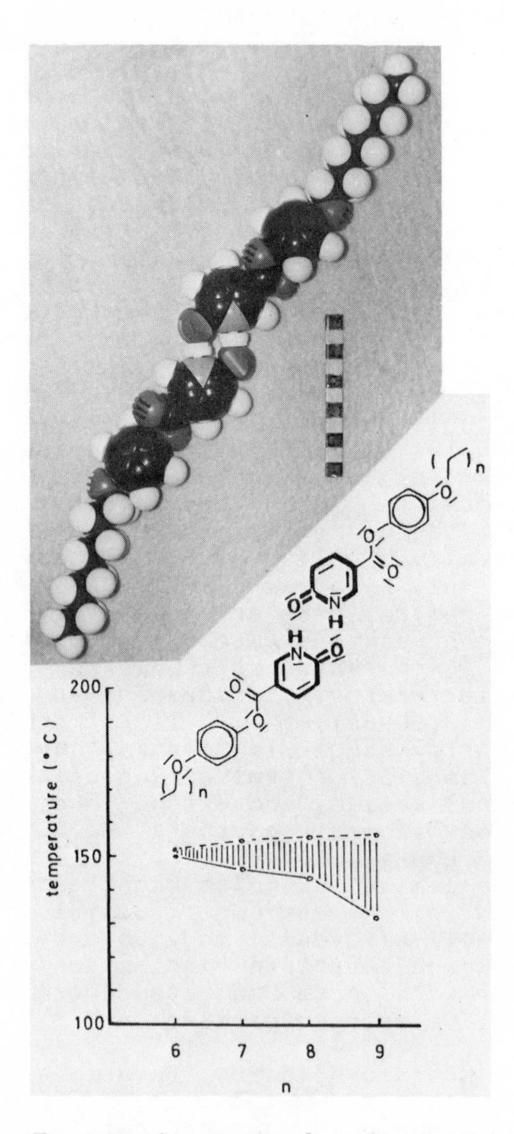

Figure 16. Smectic mesophase characteristics of 2-pyridone-5-carboxylic esters with long-chain terminals. Mesogens given in an as yet hypothetical dimer arrangement.

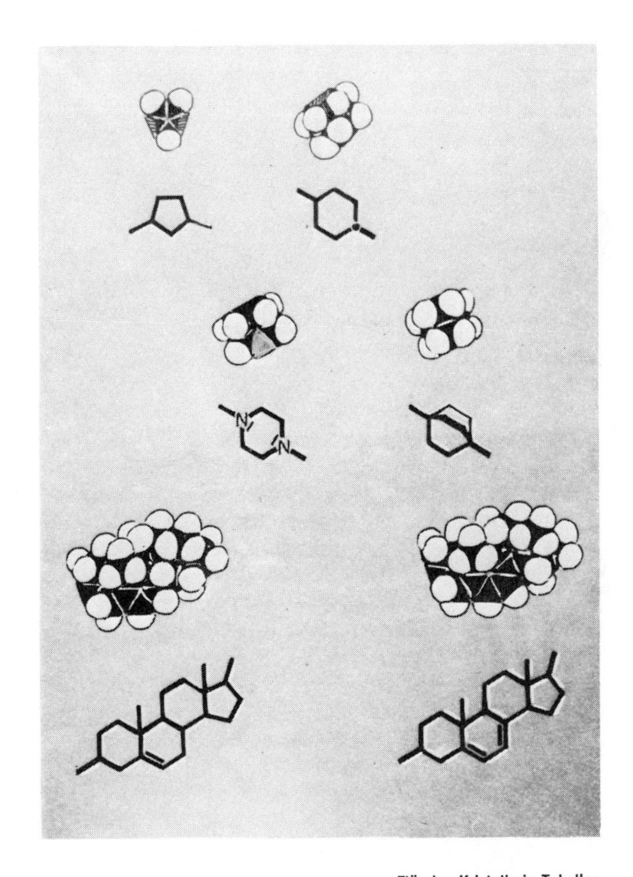

Flüssige Kristalle in Tabellen

Figure 17. Cycloaliphatic and steroidal central parts in classical thermotropics (33)

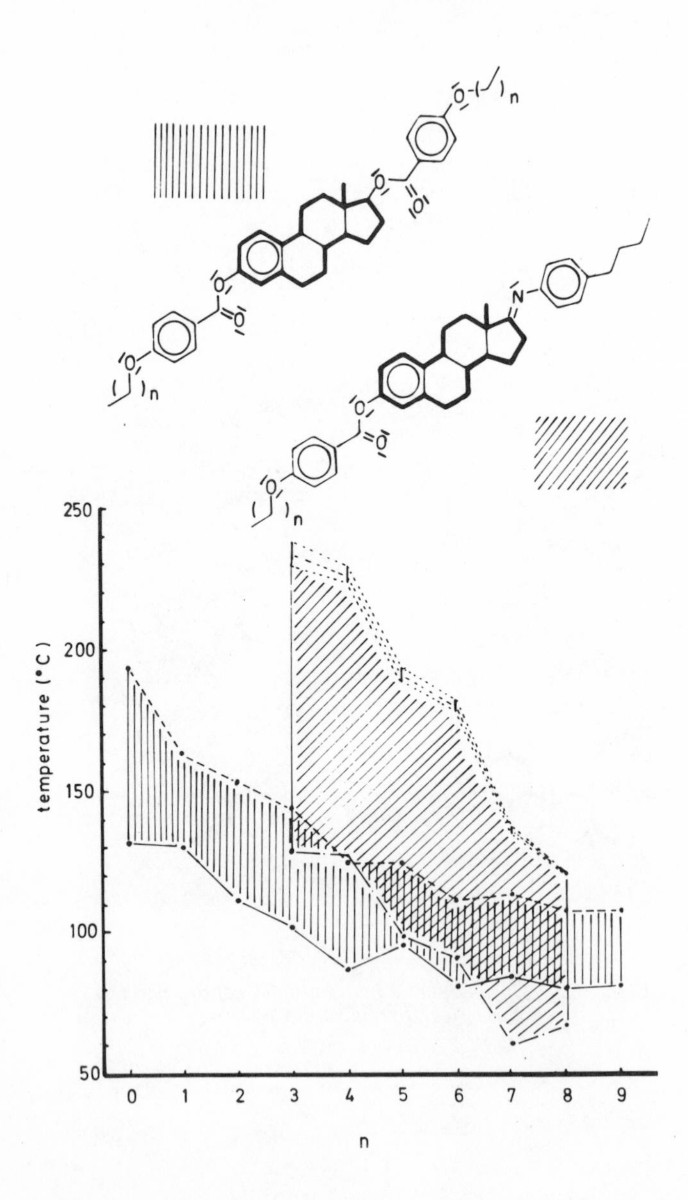

Figure 18. Mesophase characteristics of long-chain terminated estrogen cholesterics (50–55)

sierend werden kann", with those words Vorländer (107) introduced in 1913 at Halle an amplification effect, that we would like to regard as a pretty model for certain hormon like effects in bioorganisms: small amounts of an effector manipulating and directing large areas of receptor regions. About five decades after Vorländer the special interactions of small effector moieties with the base stacks of nucleic acids, dealt with in the theory of circular dichroism in nucleic acids by Tinoco and coworkers (108), stimulated Stegemeyer (109) to come back to the old Vorländer view and elucidate the special effects of those optically active doting compounds in switching mesomorphic phases of inactive materials (110-113).

Our interest in these mutual stimulations, where phase manipulations by traces of a stimulator in vitro and nucleic acid domain direction by intercalated base-pair analogues in vivo (Figure 20 and 21) (112-115) will appear as two sides of possibly common basal features, led first of all to trials to create a heuristic conception of base-pair analogues as possible effectors in manipulating the stereoelectronic pattern of nucleic acids (Figures 22, 23 and 24) (26,100). The molecular information transfer, evolved into the "channels" of specific base-base recognitions in nucleoproteinic systems should have been sensitive from its very beginning to impairments by similarly shaped molecular pattern. The evolutionary conflict with those "noising" influences might have brought about exclusions of such compounds by a refinement of protein-cover and receptor sensivities, or perhaps still more effective: – conversions and developments of those intruders into stimulating and cooperating hormonal signals. Yet an alarming abundance of our today synthetics seems to be able to break through the historically erected barriers by outwitting them with unusual, that means molecularly unreflected stereoelectronic pattern. Molecular sensitivities should be expected in competitive and allosteric sites at nucleobase-handling enzymes or at nucleic acids themselves, that is however at the essentials of our bioregulation networks.

In this context we spent some interest to the field of viro-, cancero- and immunemodulative compounds (Figures 25 and 26) (56,30,31), especially to the first monomer interferon inducer: tilorone (116-119), a fluorenone derivative, fitted with basic aliphatic terminals. As we took it as a nearly ideal example for a geometric base-pair analogue in the meaning of our conception, we postulated for this

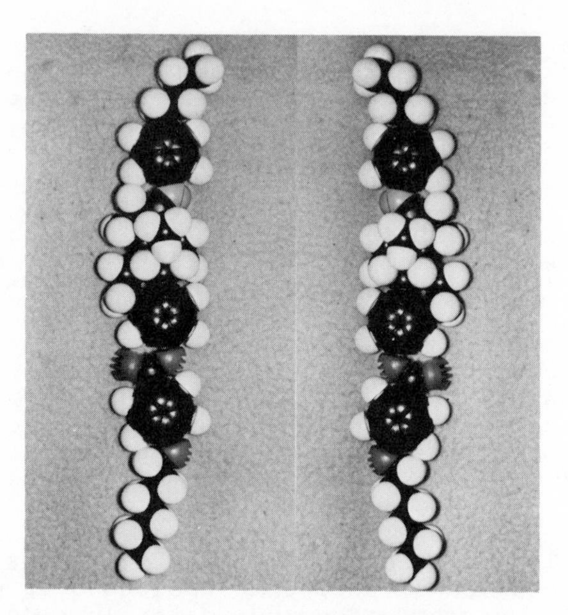

Figure 19. CPK illustration of enantiomeric long-chain terminated ester-aniles of estrone (55)

Figure 20. Switching of superheated mesomorphic pattern by intercalators in circular DNA (112, 113, 114, 115)

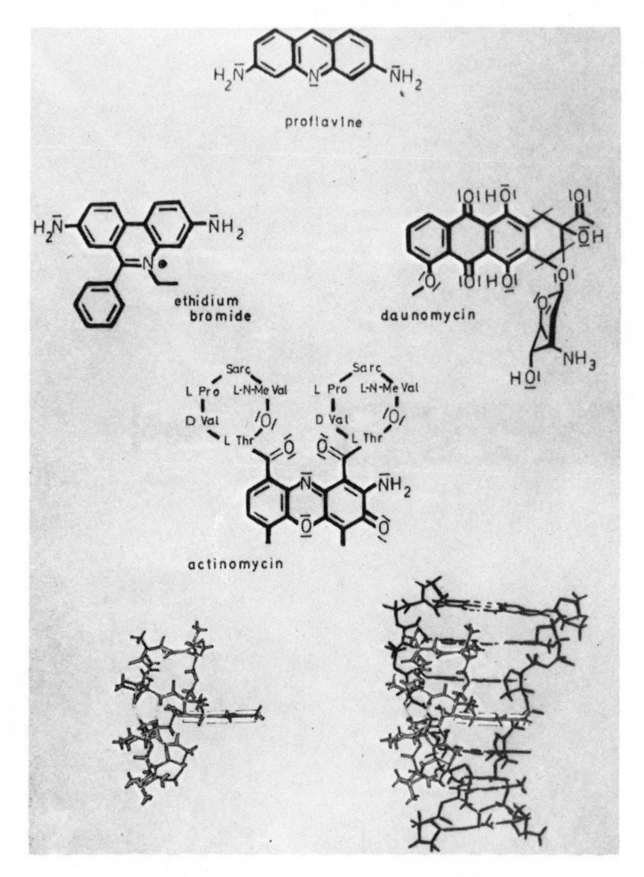

Figure 21. Examples of DNA intercalators and the Sobell visualization of the intercalation arrangement in case of actinomycin (112, 113, 114, 115)

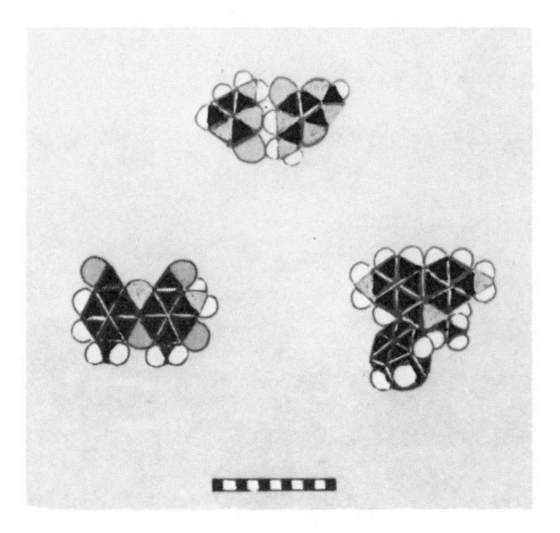

Figure 22. Watson–Crick pair in comparison with "x-ray proved" intercalator-chromophores (80, 112, 113, 114, 115)

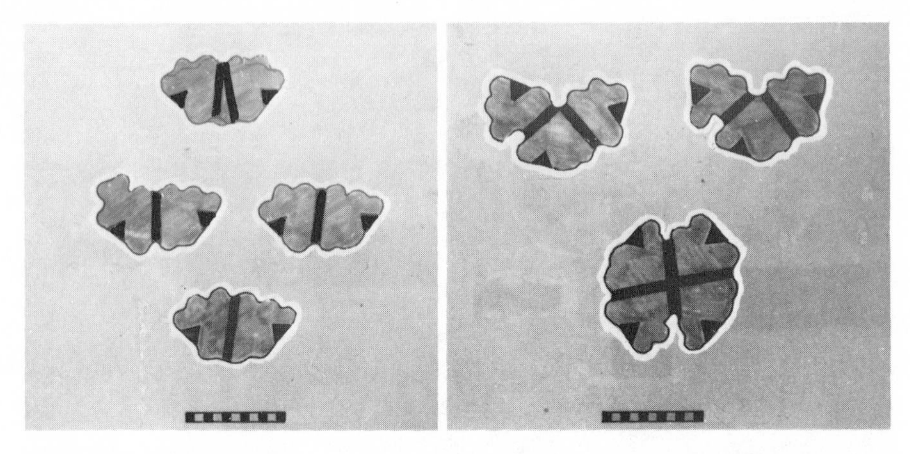

Figure 23. Overall shapes of Watson–Crick duplexes and Arnott–Bond–Felsenfeld triplexes (28, 80)

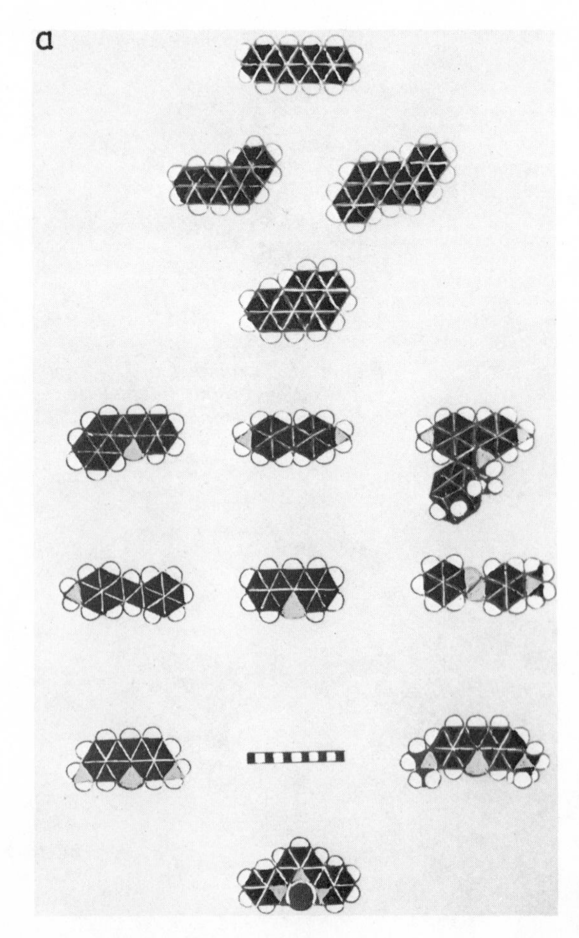

Figure 24(a). (Description on following page).

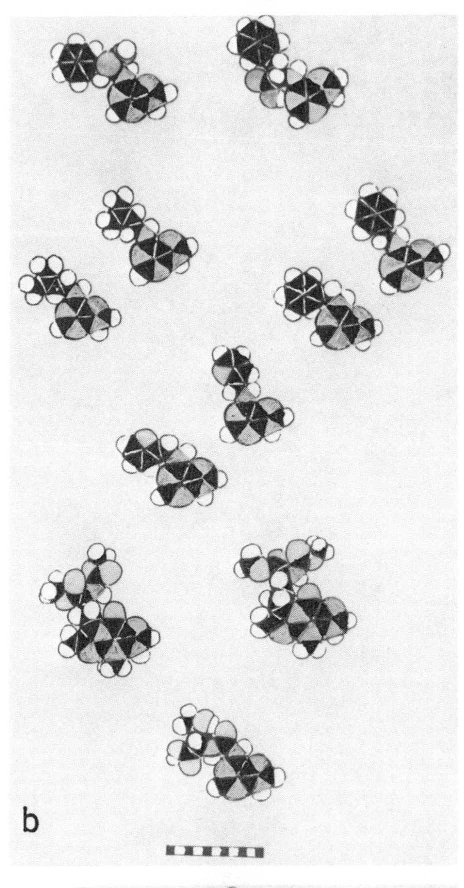

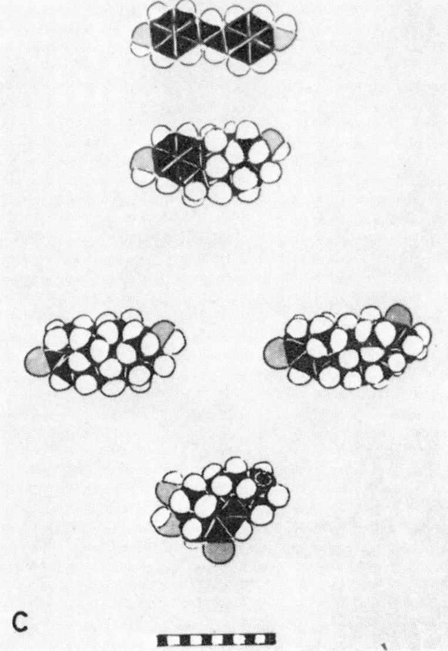

Figure 24. Illustration of "base-pair analogues" effector conception (26, 97, 98, 99, 100). Survey on simplified schemes of carbo- and heteroaromatic systems (a) (preceding page)*; cytokinines (b)* (this page, top)*; steroids (c)* (this page, bottom)*; and structural analogues and differently shaped effectors (d)* (following page) *of cellular information transfer from nucleic acids to protein.*

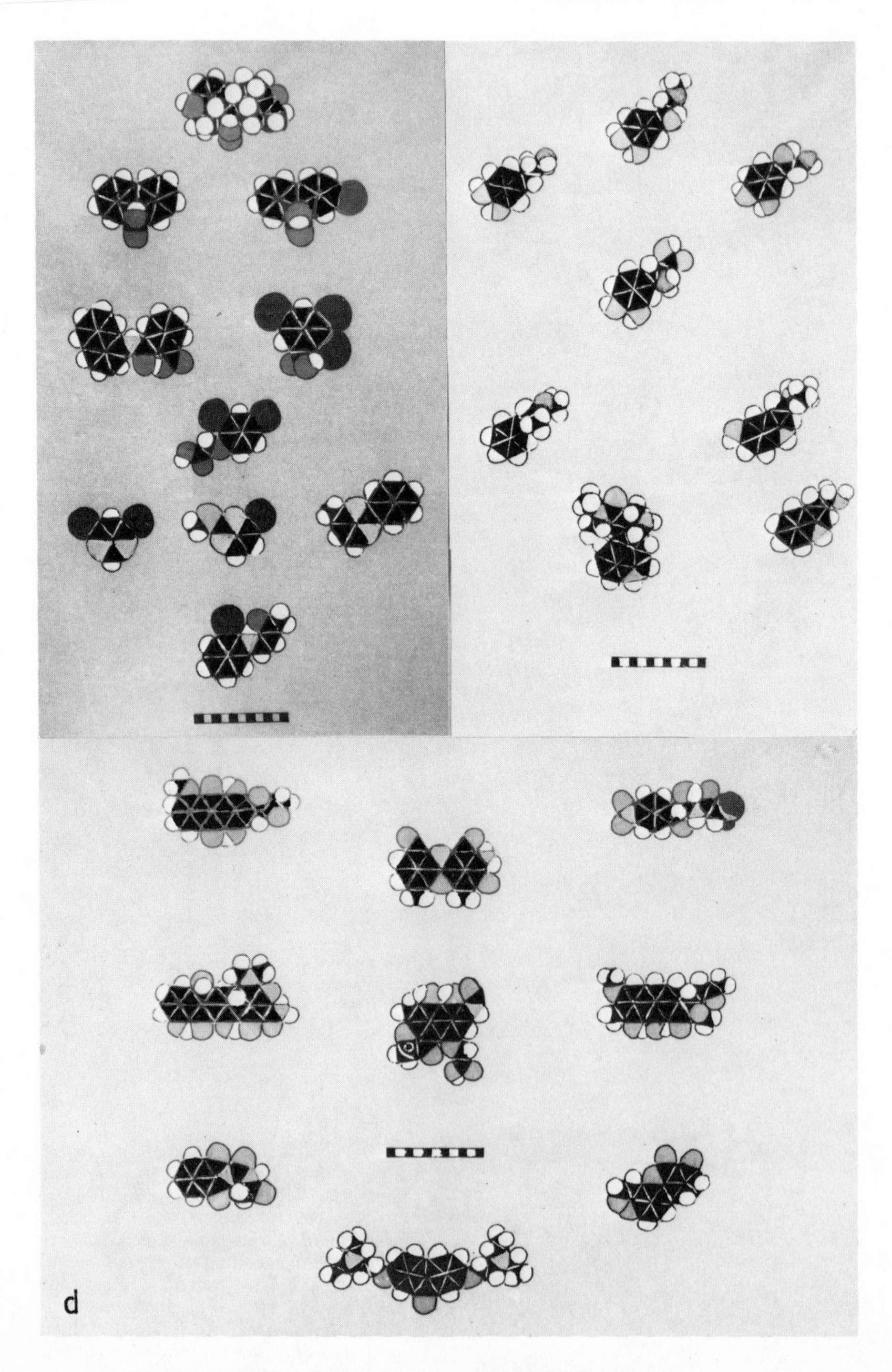

Figure 24(d). (Description on preceding page).

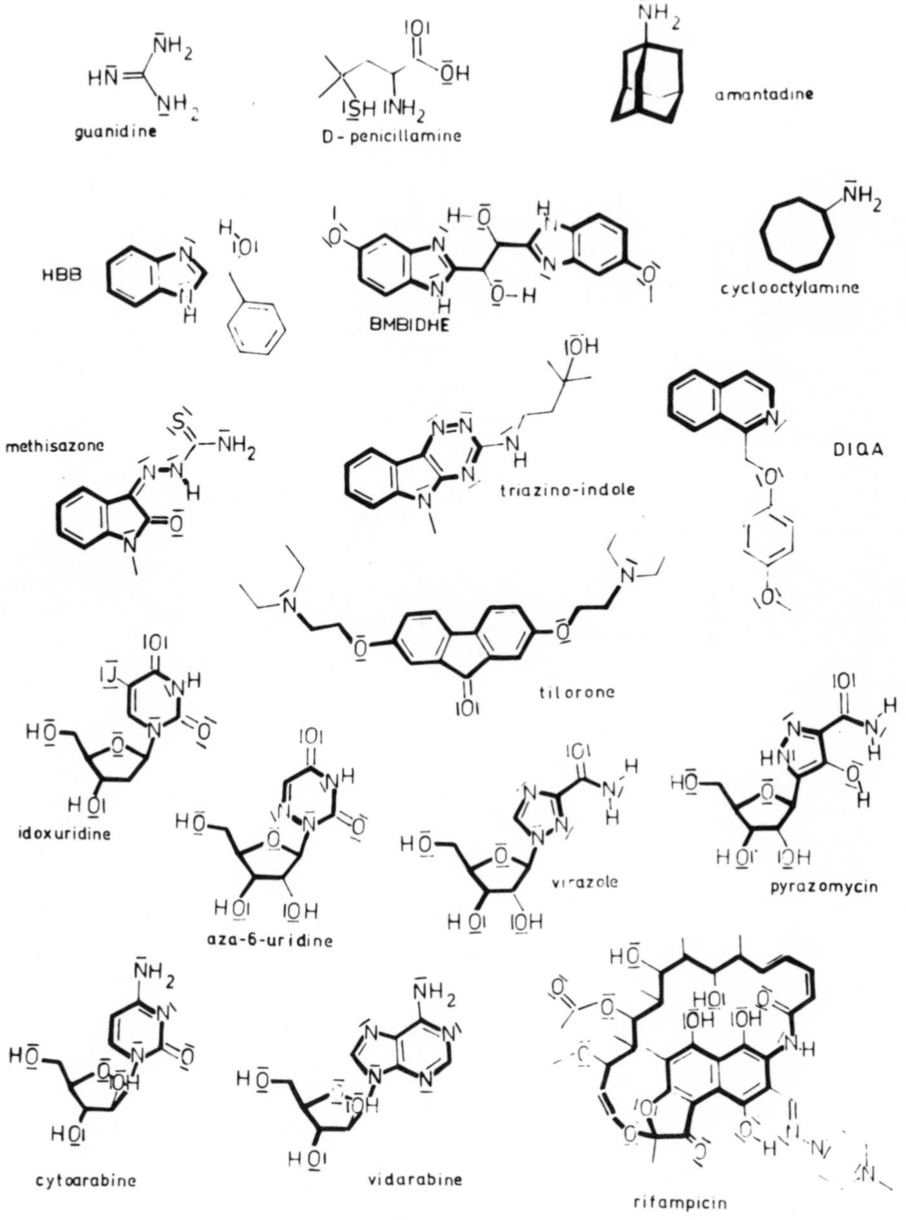

guanidine

D-penicillamine

amantadine

HBB

BMBIDHE

cyclooctylamine

methisazone

triazino-indole

DIQA

tilorone

idoxuridine

aza-6-uridine

virazole

pyrazomycin

cytoarabine

vidarabine

rifampicin

Figure 25. "Monomeric" antivirals

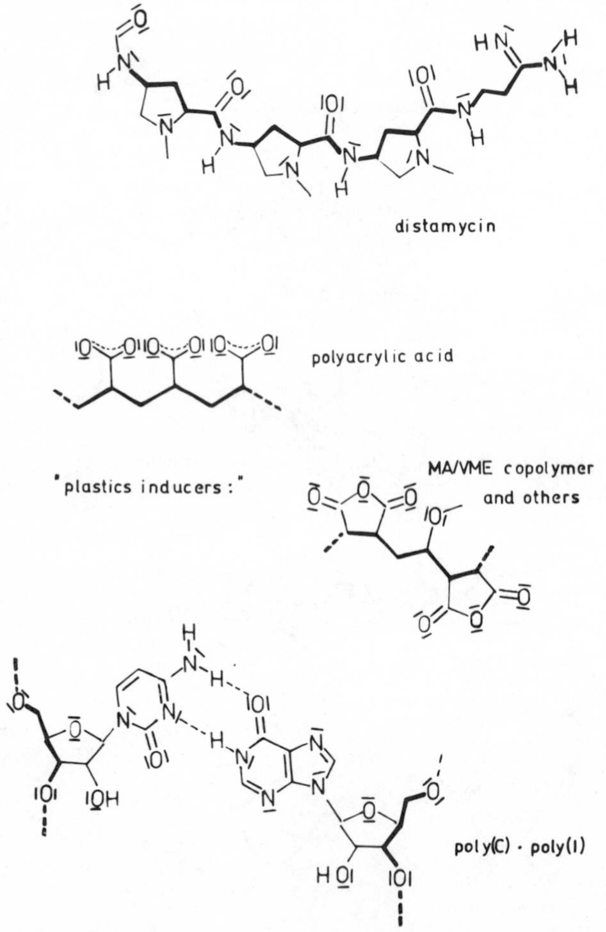

distamycin

polyacrylic acid

"plastics inducers:"

MA/VME copolymer
and others

poly(C) · poly(I)

Figure 26. "Oligomeric and polymeric" antivirals

compound an intercalation mechanism into DNA (100), which has been proved later on by Chandra and coworkers (118). Independently in 1976 an Austrian group (120) and we started a stereoelectronic simulation program (77) (Figure 27), concentrating our interest in the substitution of the fluorenone ether parts by carbonic acid amido groups. Because of the relationships of the latter to actinomycin, distamycin and some other similarly designed intercalators (115,100, 98), we hoped to get some additional insight into the vertical and horizontal, the stacking and hydrogen bonding activities of those effectors. For the most directly modelled tilorone analogue, the fluoramide FA-1 (Figure 27) - common to the synthetical efforts of both groups - the Austrians and we proved antiviral in vivo effects in mice in general and interferon induction potencies in particular. But mainly interested in interaction possibilities of these structures with nucleic acids - linking the in vivo and in vitro aspects of the old Vorländer-Stegemeyer story - we can, in addition to the well known characteristics of fluorenone pattern to display classical thermotropic behaviour (33) (Figure 28), present further evidence for its capability in inducing strandreorientations by attaching, stacking and intercalation interactions with DNA (77), indicative by the shift of the melting point in the presence of the intercalator and by the induced circular dichroism on titrating DNA with a solution of the effector molecule (Figure 29). Figure 30 tries to give a hypothetical look on the fit of the molecule to DNA: starting with an electrostatically directed attack into the large groove and reaching a considerable number of favorable contacts on intercalation, with the aromatics sandwiching one another, phosphate and amino groups fixing the aliphatic terminals and perhaps hydrogen bonds linking the carbonamide groups to the 1'-oxygens of the ribose moieties of both strands, the whole mechanism drives the strand arrangements of the "infected" B-DNA into the direction of an RNA thus sensitizing this parts for opposite receptors (28,98). Aspects of specifity to special base regions - at present under investigation in our laboratory - seem to reveal a slightly different picture compared with tilorone as to the preference of AT-, instead of GC-pairs (77,116-119).

In this way the old Vorländer effect of directing mesophase systems by interaction possibilities of small amounts of inducers, elucidated and used in theoretical and practical aspects by Stegemeyer and his group mainly in the field of classical thermotropics,

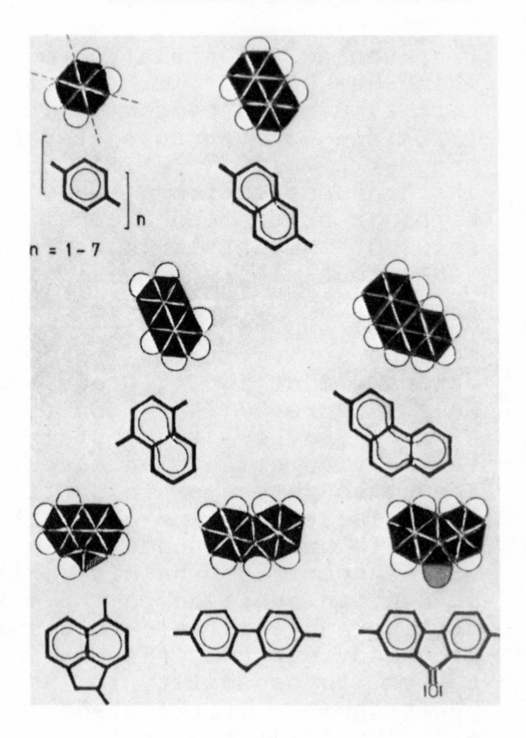

Flüssige Kristalle in Tabellen

Figure 27. Fluorenes and fluorenones as central parts of classical thermotropics (33)

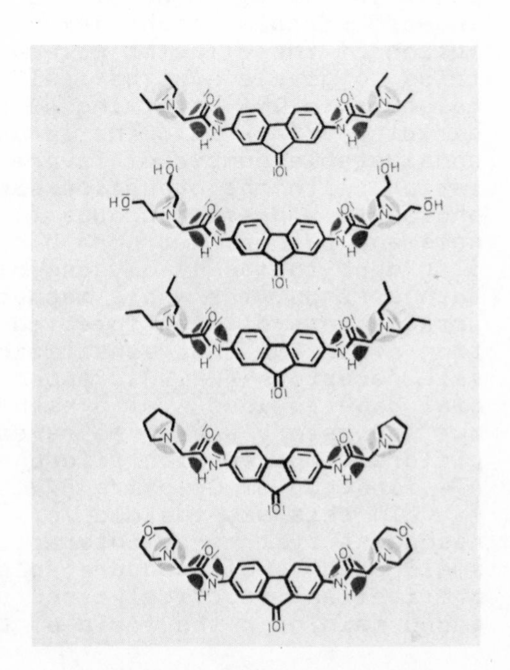

Figure 28. Fluoramides FA-1–FA-5 as stereoelectronic simulations of tilorone (56, 57)

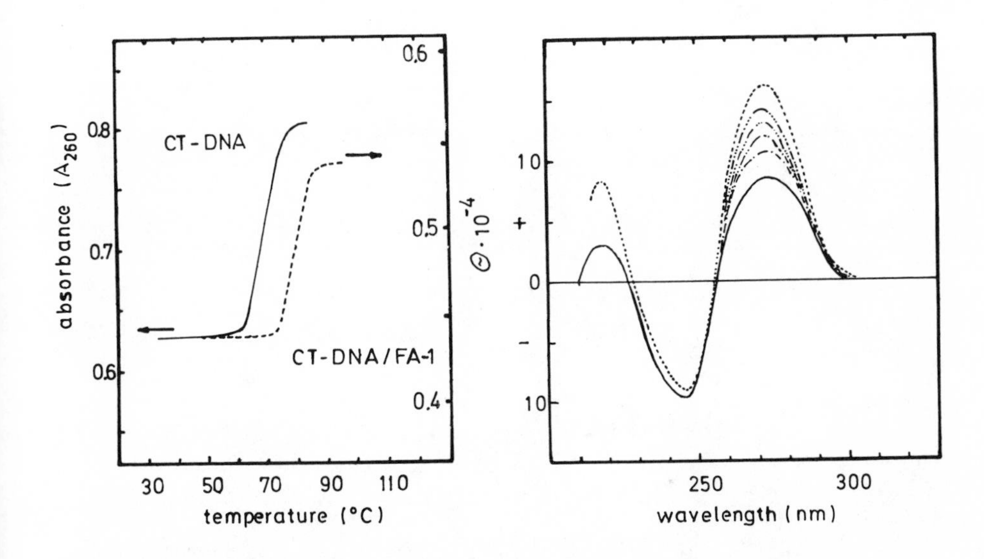

Figure 29. Thermal melting profiles and ORD-characteristics of CT-DNA/FA-1 interactions (76, 77). (a) Melting profiles of CT-DNA (OD_{260} 0.625, 0.001M NaCl, pH ~ 7, compared with the melting characteristics of CT-DNA/FA-1 · 2HCl complexation. (3 · 10^{-3}M aqueous solution of FA-1 · 2HCl; ratio DNA-phosphate/FA-1 = 1/1). (b) OD-characteristics of CT-DNA/FA-1 · 2HCl interactions on titration. CT-DNA OD_{260} 0.902, 0.1M NaCl; other conditions like (a).

Curve	Ratio DNA-P/FA-1 · 2HCl
—·—	*1/0.1*
—··—	*1/0.15*
—···—	*1/0.2*
—····—	*1/0.3*
----	*1/0.5*

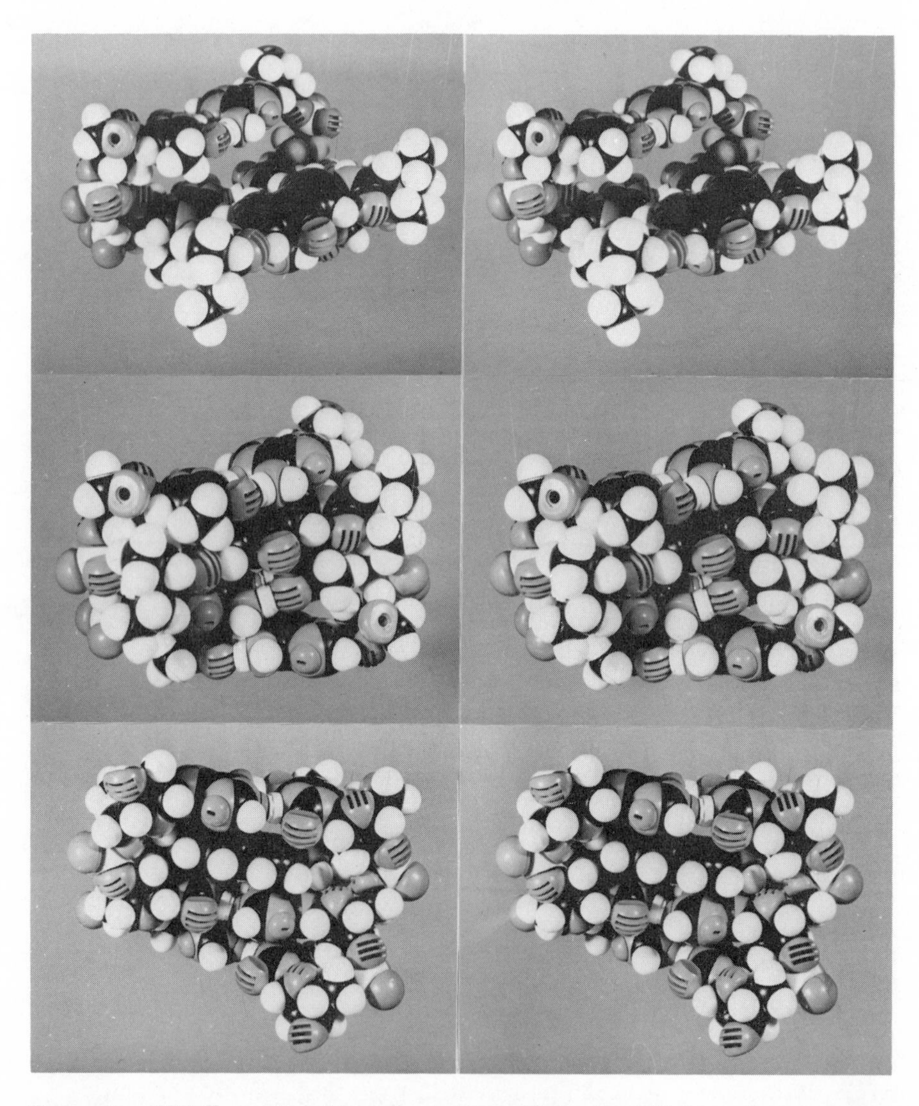

Figure 30. CPK-illustration of hypothetical intercalative fit of FA-1 into the large groove of DNA. (from top to bottom) *FA-1 approaching the major groove of DNA; intercalative fit from the major groove; the same arrangement from the minor groove* (56, 76, 77).

brought back to their theoretical stimulus seems to gain a new dimensions for modelling hormone and effector efficiencies in general and intercalator effects in the nucleic acid base stacks in particular. It offers a possible explanation for the viro-, cancero- and immunemodulative activities of those compounds in their ability to strangen the common nucleic acid pattern of bioorganism and to provoke both humoral and cellular non-self recognitions, while the old Vorländer term of "infection" - rather unusual in the field of thermotropics - fits our today biological aspects in a curious way, somewhat indicative and characteristic for the inherent connection.

Modelling and "Strangening" Nucleic Acids by Polynucleotide Strand-Analogues

Base and sugar variants of nucleosides and nucleotides have been dealt with in the past in an abundance, nearly incapable for reviews, monographs or even reference journals (26,121-123). While these variants of a first and second generation of analogues still continue attracting the attention of people interested in nucleic acid chemistry and its applications (123), beginning with the 1960s a third generation of nucleic acid simulations emerged from the flood of stereoelectronic imitations: far or less distantly modelled strand analogues revealed some importance for both theoretically and practically oriented investigations in the field of biopolymers and their interaction possibilities (26,58-61,97,124-131).

To deal only with those structures, modelled from a greater distance, there opened for synthetical approaches ways of polycondensation, polyaddition free-radical, photochemical, or γ-ray induced polymerizations or polymeranalogous reactions, possibilities that have been met in the meantime by numerous synthetic efforts, reviewed already by different authors (26,59,97,128-130,132,133).

As a compromise between the screening postulations of rapid and uncomplicated access to high molecular systems (due to the selections of a size criterion for interaction possibilities of biopolymers within bioregulation hierarchies (134)) by a route independent from the natural polycondensation way (due to a greater suitability for the alternatives of matrix polymerizations) and a certain degree of stereoelectronic order of our synthetics (due to the desire for reproducibility and fit of our systems in their possible interactions with biopolymers) we started from N-vinylnucleobases (26,97,125), that have also been used by American and Japanese groups (58-61,130, 133). However, differing from their general conceptions, we have tried from the beginning to fit minimally altered base stacks by variable strand patterns, thus to overcome the otherwise obvious limitations for useful variations in strand-pattern design (64,135).

As model building provided poly(N-vinylnucleobase-vinylX)-systems of alternating sequence and syndiotactic (isotactic respectively) stereo (isotactic in 1.5-base substituents) being capable of polynucleotide-like hydrogen-bonding and base-stacking actions at least over shorter distances (99,125,135), we intended to favor our synthesis conditions into the direction of this working hypothesis. But as we felt ourselves unable to elucidate and grant those very conditions, we hoped, that our systems would at least partially meet of intensions half-way. As nucleobases in aqueous solutions usually exhibit limited numbers of repeating pattern of stacking interactions (136, 137), we suggested that certain preorientations - remaining even under our unfavourable synthesis conditions - could build up stack matrices, that might practice some sort of sequence- and stereo-regulation. In these context the following Figures 31-36 display in a though intriguing nevertheless mere hypothetical way our hopes and expectations, starting with the inviting tendencies of vinyl polymers for helical arrangements (138,139) (Figure 31) over the useful stacking ambitions of nucleobases, given here for example for different adeninestacks (Figure 32) (135) and one vertical interaction of a hypoxanthine moiety (Figure 33) (140), to the speculations of vinyl nucleobase homo- and copolymerization arrangements at the growing-up chain in the scheme of Figure 34, illustrated further in case of 9-vinylhypoxanthine-acrylic acid copolymerization (Figures 35 and 36) and ending with the partial $(I)_n \cdot (C)_n$-simulation as an example for possible biological activities (30).

Varying both base and strand compositions, a spectrum of polymers - idealized with regard to ratio, sequence and structure fit in Figures 37 and 38 - were prepared by AIBN- or γ-ray induced copolymerizations, followed in case of $(vA,[vOH]_{0.7})_n$ by a polymeranalogous reaction. The starting amounts of N-vinylnucleobases and comonomers were adjusted to yield a nearly 1:1 ratio in the polymers. Figure 39 adds for some characteristic cases the received S-value distributions, giving an idea of the molecular weight distributions (66).

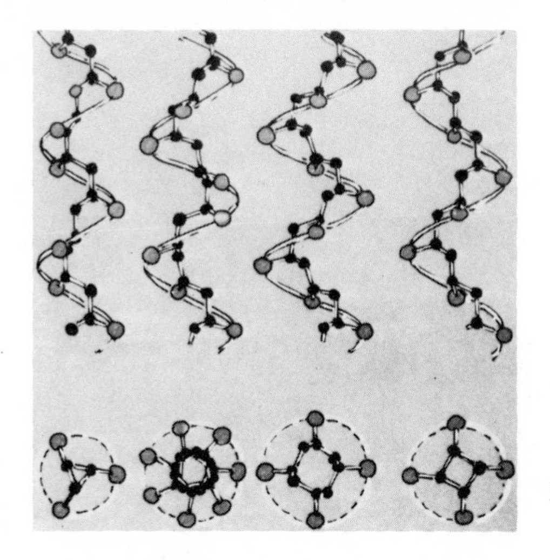

Introduction to Stereochemistry

Figure 31. Helical ambitions of simple vinyl polymers after (138)

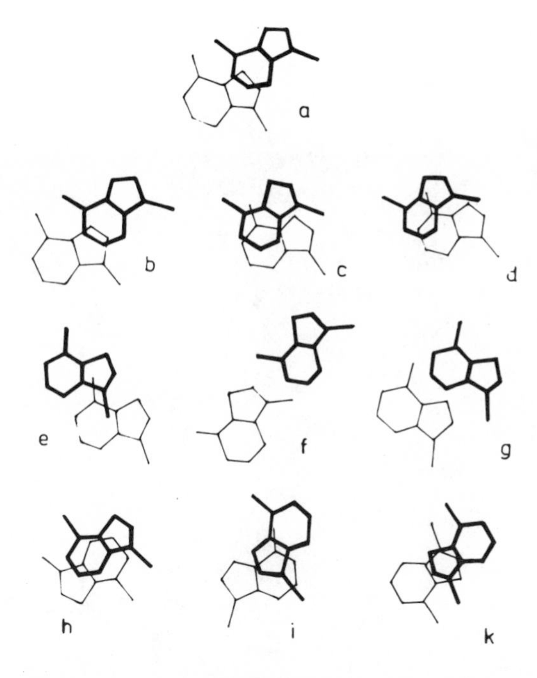

Figure 32. Adenine-stacks after (136, 137). (a) RNA; (b) A-DNA; (c) B-DNA; (d) C-DNA; (e) deoxyadenosine-monohydrate and adenosine-5′-phosphate; (f) adenosine-3′-phosphate; (g) adenine-hydrochloride and 9-methyl-adenine-dihydrobromide; (h/k) 9-methyl-adenine; (i) $(A)_n$.

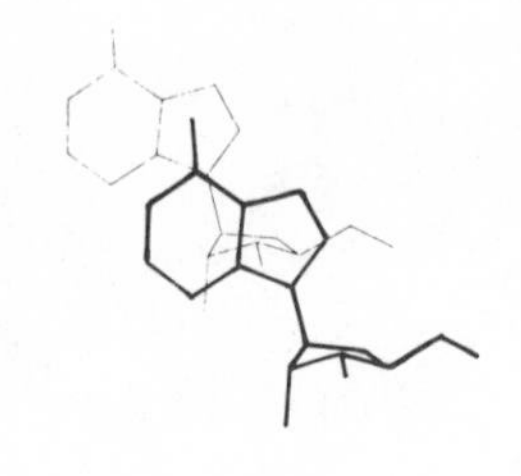

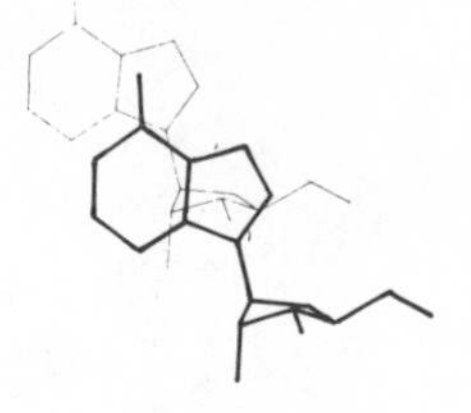

Journal of Molecular Biology

Figure 33. Inosine stacks after (140)

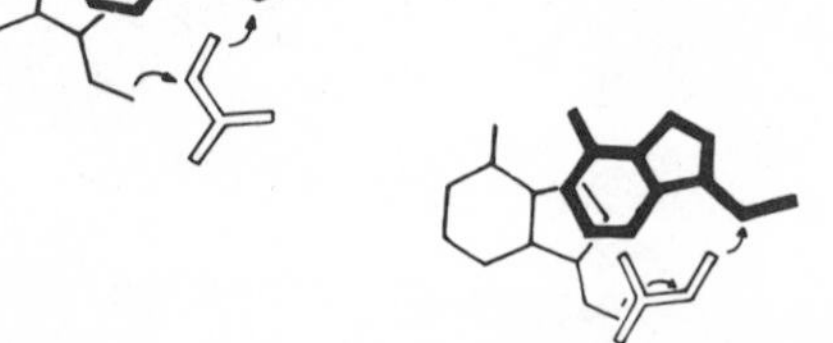

Figure 34. Hypothetical insertion of vinyl comonomers into vinyl nucleobase-stack matrices, bridging gaps between the vinyl groups and allowing a convenient zipping-up (26, 64, 125, 135)

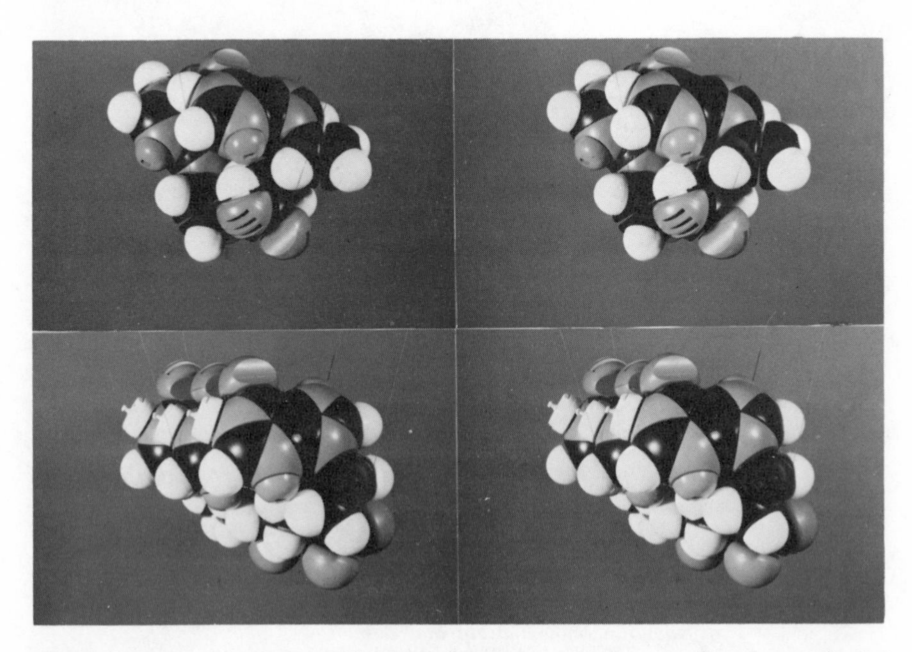

Figure 35. CPK illustration of the speculation given in the preceding figure. (top) *9-Vinylhypoxanthine/acrylic acid-stack insertion-arrangement;* (bottom) *built up (vH-vCOOH)*$_n$.

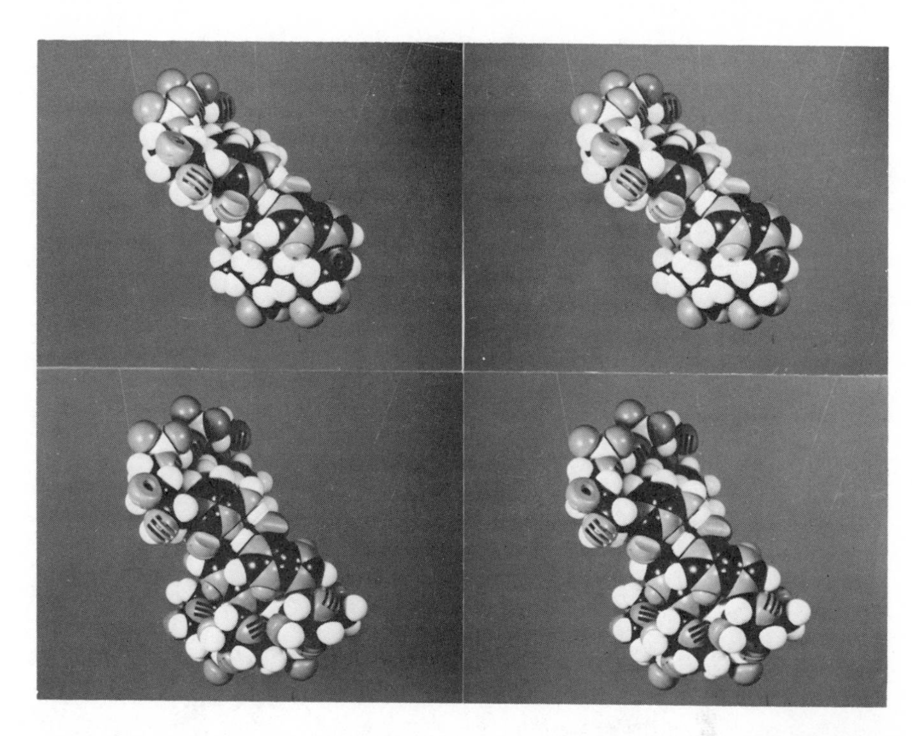

Zeitschrift fuer Chemie

Figure 36. CPK illustration of the hypothetical $(C)_n \cdot (vH,vCOO^-)_n$ hybrid compared with the all-polynucleotide base-paired $(C)_n \cdot (I)_n$, vizualizing the hydrogen-bonding and stack similarities between the polynucleotide/mutual strand-analogue hybrid and the polynucleotide duplex (62)

Figure 37. Idealized schemes of vinyl copolymers with varied nucleobase design (top to bottom and left to right): $(vA\text{-}vCOONa)_n$; $(vU\text{-}vCOONa)_n$; $(vC\text{-}vCOONa)$; $(vH\text{-}vCOONa)_n$ (64–71, 75)

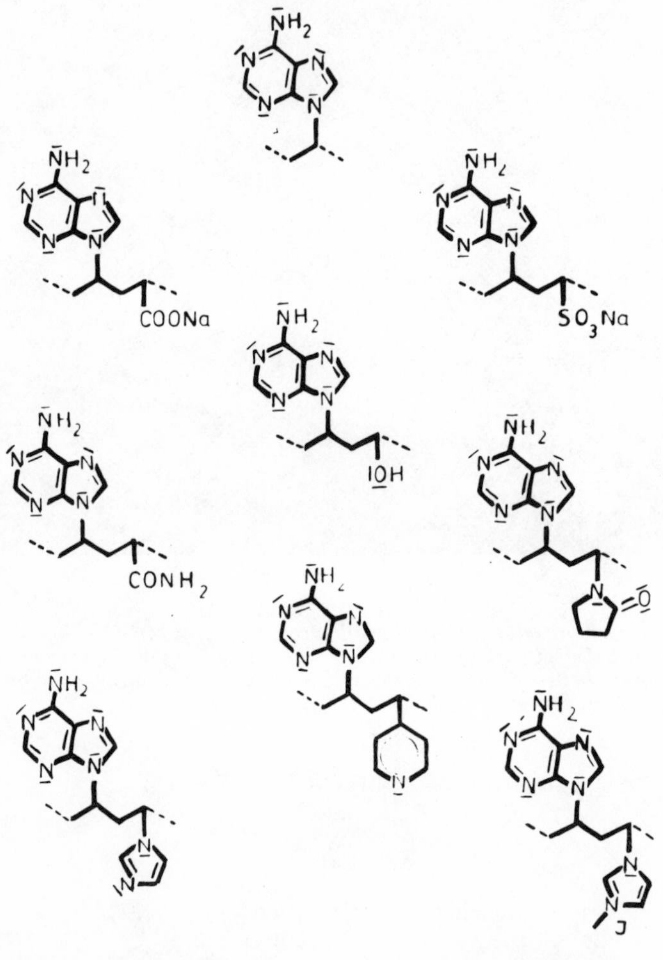

NH2
COONa
SO3Na
OH
CONH2
O
N
J

Figure 38. Idealized schemes of 9-vinyladenine copolymers with varied strand polarity pattern (top to bottom and left to right)*: $(vA)_n$; $(vA\text{-}vCOONa)_n$; $(vA\text{-}vSO_3Na)_n$; $(vA\text{-}vOH)_n$; $(vA\text{-}vCONH_2)_n$; $(vA\text{-}vPn)_n$; $(vA\text{-}vP)_n$; $(vA\text{-}vI)_n$; $(vA\text{-}vm^3IJ)_n$* (64–71, 75)

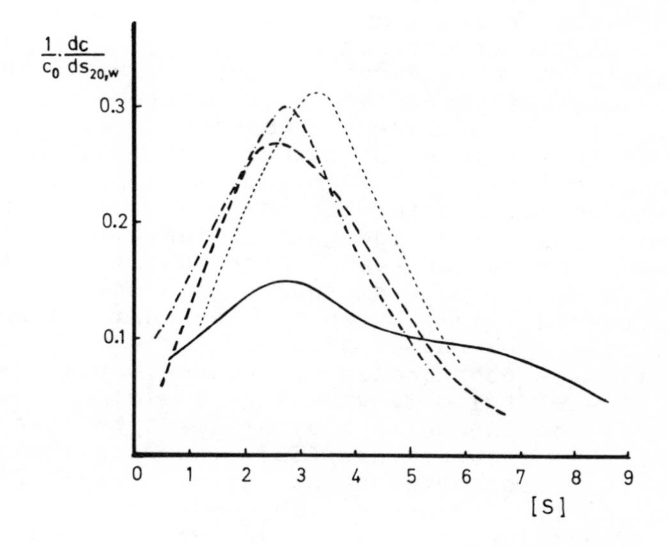

Zeitschrift fuer Chemie

Figure 39. S-value distribution of some characteristic copolymers. (——) (vA, [vOH]$_{0.7}$)$_n$; (---) (vA, [vCOONa]$_{0.8}$)$_n$; (—·—) (vA, [vCOONa]$_{1.4}$)$_n$; (- - -) (vA, [vP]$_{0.9}$)$_n$ (66).

To check our thoughts as to the possibilities for polynucleotide-like mesomophic order in our synthetic products, base-pairing experiments with their mate polynucleotides have been carried out including the whole set of usual complementary complex formations (64,68,73,74,97). The investigations - illustrated by mixing curves, thermal melting profiles, and ORD-as well as CD-characteristics of Figures 40 and 41 - reveal that cooperativity within the "base-paired complexes" varies over a rather broad range, reaching the qualities of mutual polynucleotide interactions in case of the rather vA-rich $(vA,[vOH]_{0.7})_n$ hybridized with the poorly structured $(U)_n$ (141) - demonstrated by definite and considerable hypochromism in the mixing curves, sharp absorbance temperature profile and individually shaped ORD- and CD-spectra -, and dropping down to the level of rather limited interactions by chance in case of the numerous more poorly fitted analogues in combination with highly ordered, self-associated polynucleotides, indicated by low and diffuse hypochromism (or even hyperchromism), flat melting profiles and only disturbed polynucleotide-ORD- and CD-spectra. As we up to now have not got any possibility for an x-ray proof of our original hopes and expectations - represented by the early picture of Figure 42 - we tried to visualize one of our hybrides (of course not the worst one) at least by electron microscopy as a substitute (65), and here the overall shape simulations between the "native" $2(U)_n \cdot (A)_n$ and the semiplastic $2(U)_n \cdot (vA,[vOH]_{0.7})_n$ seems to add further evidence for structural similarities and accomplishes comparable results of the Pitha group (130) in the field of homopolymers (Figure 43).

From all of this the conclusions might be drawn very carefully, that our hypothesis works best - if at all - in the polymerization of highly stacking vinyl nucleobases, favoring perhaps at least in these cases some sort of combined stereo- and sequence regulation by self-organized stack matrices, and ordering the expected jungle of sequence and stereoirregularities (at present we have neither really convincing indications for the prefer ence of the expected sequence alternation nor for the less favorable but possible alternative of block polymer formation) at least partially for short interplays of monotonic sequences of polynucleotides with limited recognition tracts of our synthetics - allowing the curious games of acquaintances between late derivatives of evolutionary optimizations and pattern, both relatives of ancestors of early nucleic acids ambition in nature

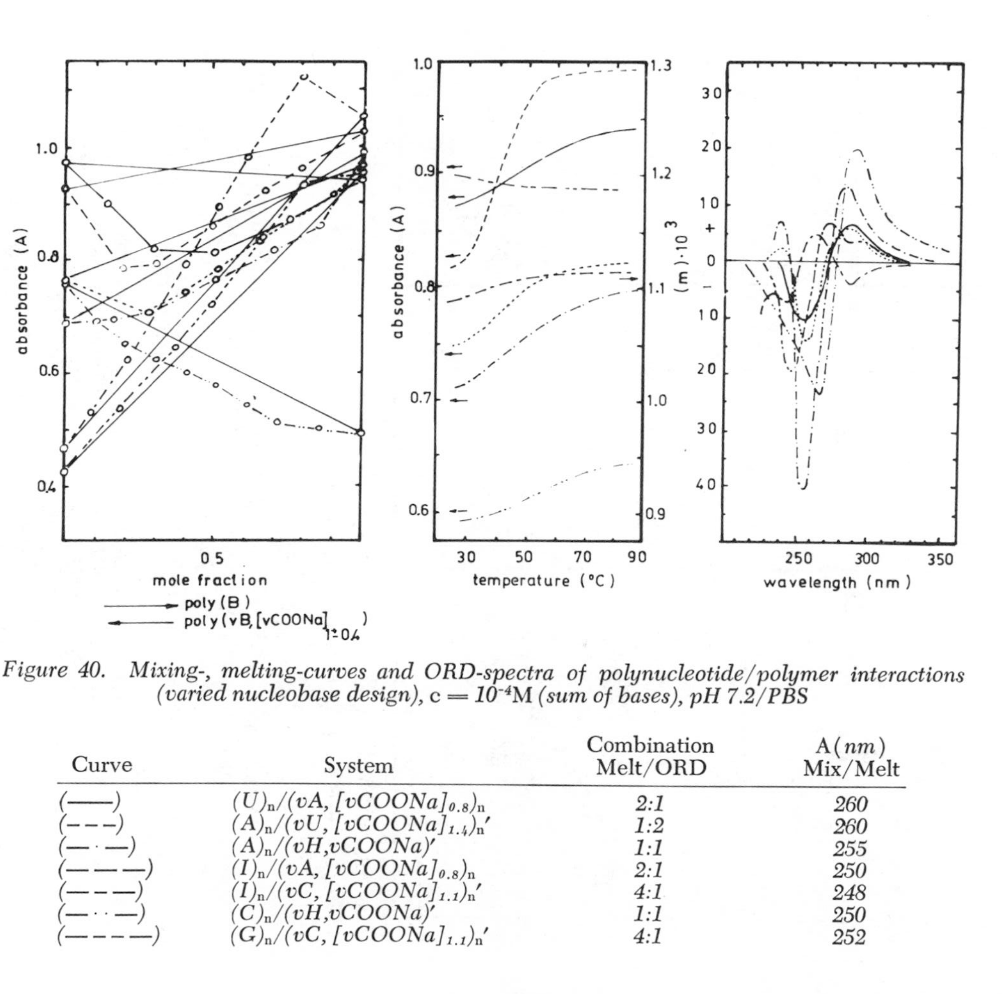

Figure 40. Mixing-, melting-curves and ORD-spectra of polynucleotide/polymer interactions (varied nucleobase design), c = 10^{-4}M *(sum of bases), pH 7.2/PBS*

Curve	System	Combination Melt/ORD	A(*nm*) Mix/Melt
(——)	$(U)_n/(vA, [vCOONa]_{0.8})_n$	2:1	260
(- - -)	$(A)_n/(vU, [vCOONa]_{1.4})_n'$	1:2	260
(—·—)	$(A)_n/(vH, vCOONa)'$	1:1	255
(— — —)	$(I)_n/(vA, [vCOONa]_{0.8})_n$	2:1	250
(— - —)	$(I)_n/(vC, [vCOONa]_{1.1})_n'$	4:1	248
(—··—)	$(C)_n/(vH, vCOONa)'$	1:1	250
(— - - - —)	$(G)_n/(vC, [vCOONa]_{1.1})_n'$	4:1	252

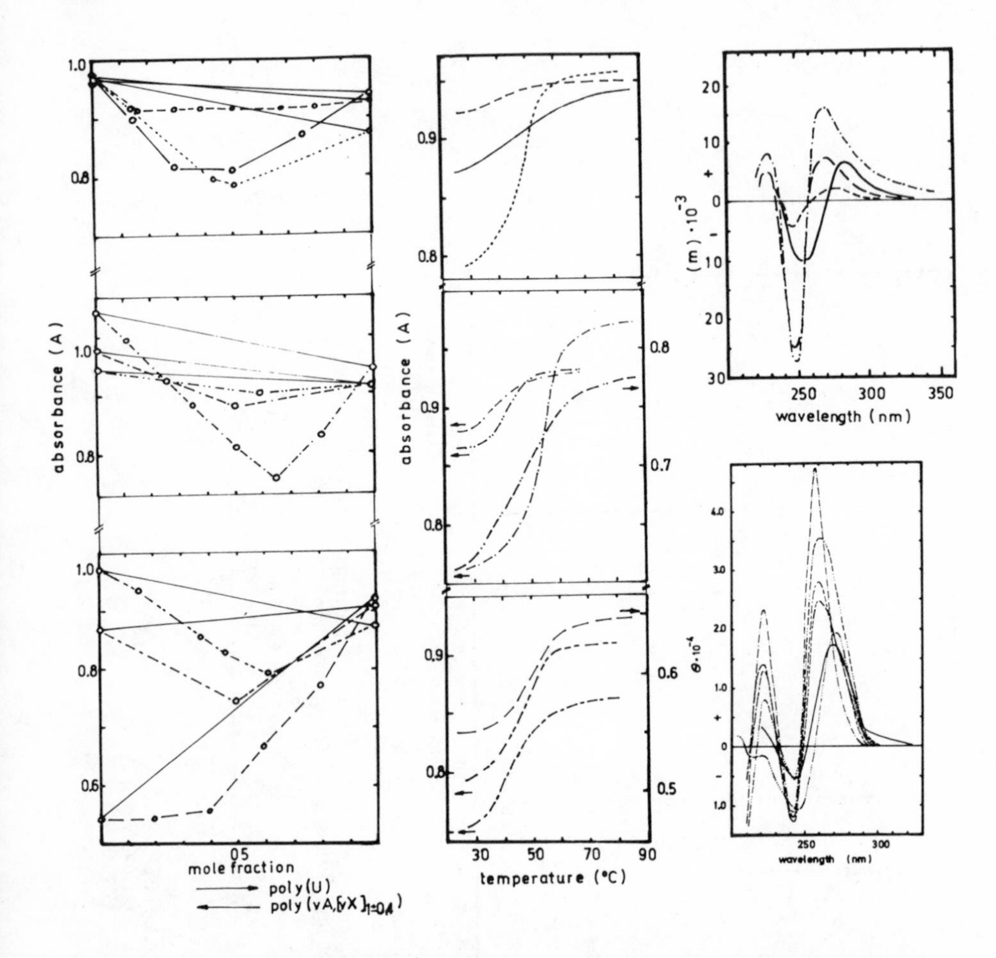

Figure 41. Mixing-, melting curves and ORD-spectra of polynucleotide/polymer interactions (varied strandpolarity pattern), c = 10^{-4} M (sum of bases), pH 7.2/PBS

Curve	System	Combination Melt/ORD/CD	A(*nm*) Mix/Melt
(——)	$(U)_n/(vA, [vCOONa]_{0.8})_n$	2:1/2:1/2:1	260
(– – –)	$(U)_n/(vA, [vCOONa]_{1.4})_n'$	1:4/1:4/ —	258
(- - -)	$(U)_n/(vA, [vSO_3Na]_{0.7})_n$	3:2/ —/ —	255
(—·—)	$(U)_n/(vA, [vOH]_{0.7})_n$	2:1/2:1/2:1	260
(—··—)	$(U)_n/(vA, [vCONH_2]_{1.1})_n$	3:2/ —/ —	258
(——·)	$(U)_n/(vA, [vPn]_{0.8})_n$	1:1/ —/ —	258
(———)	$(U)_n/(vA, [vP]_{0.9})_n$	1:2/1:2/ —	258
(———)	$(U)_n/(vA, [vI]_{1.1})_n$	1:1/ —/1:1	257
(————)	$(U)_n/(vA, [vm^3I]_{0.6})_n$	3:2/ —/1:1	255
	for comparison:		
(——·——)	$(U)_n/(A)_n$	1:1/ —/ —	258
(———···)	$(U)_n/(A)_n$	—/ —/2:1	—
(—····—)	$(U)_n$	—/ —/1	—

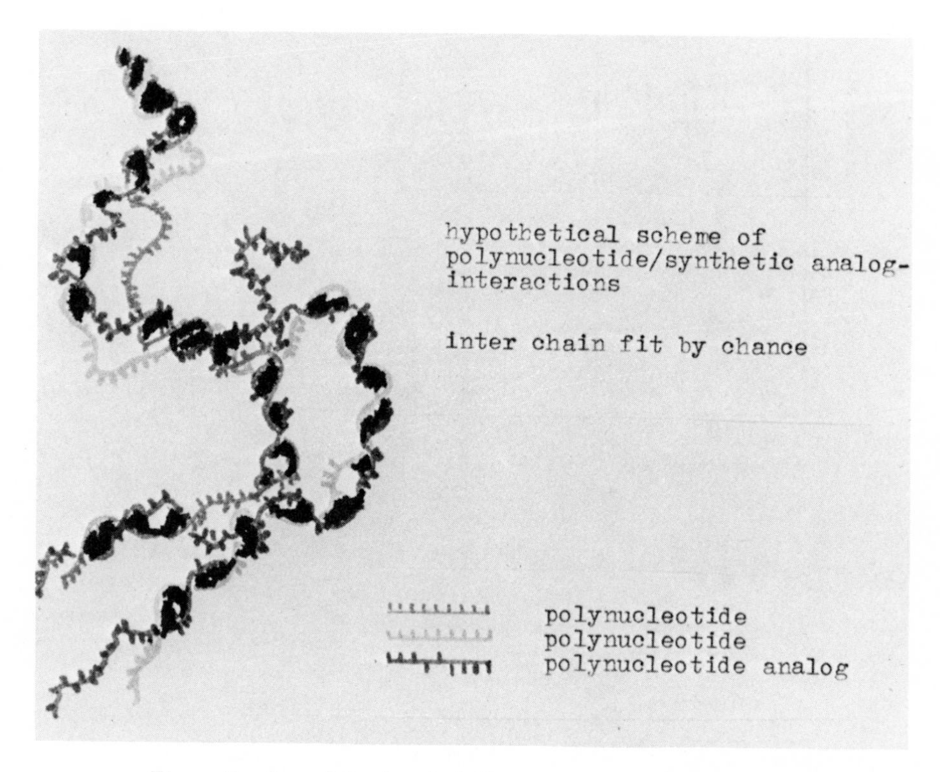

Figure 42. Speculative interhybrid arrangements and interactions

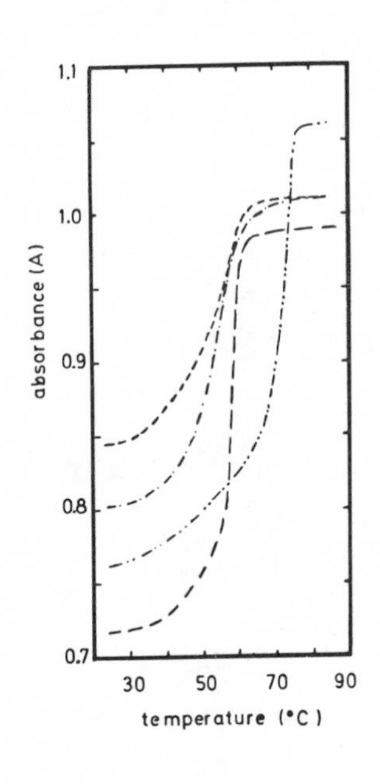

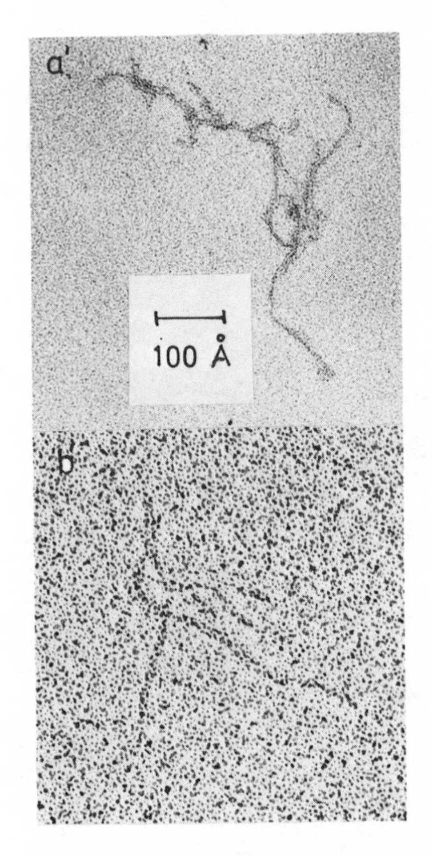

Zeitschrift fuer Chemie

Figure 43. Electron micrographs of $(U)_n \cdot (A)_n \cdot (U)_n$-triplex and its partial analogue with additional interpretation of complexation by melting profiles (65).

(a) PBS, pH 7.2:

(— — —) $2(U)_n \cdot (A)_n$ A_{255}

(— · —) $2(U)_n \cdot (vA, [vOH]_{0.7})_n$ A_{260}

(b) 1M NH_4 ac.; 8 mM NaCl; 0.5 mM phosphate; pH 7.1:

(— ·· —) $(U)_n \cdot (A)_n$ A_{258}

(- - -) $(U)_n \cdot (vA, [vOH]_{0.7})_n$ A_{258}

(a′) $2(U)_n \cdot (A)_n$—$UO_2(OAc)_2$-stained; (b′) $2(U)_n \cdot (vA, [vOH]_{0.7})_n$—Pt-stained solution of a′ and b′ like b

and creations of late nucleic acid ambitions in our brains.

Though all these hypotheses have to be looked upon with caution, since we can only supply indirect evidence in favor of it, according to the indications from these base-pairing experiments, our synthetic polymers seem to reveal both structural and functional similarities with their polynucleotide standards. Thus one might hope or speculate, that after a period of necessary stereoelectronic optimizations - their "native" ideals "wasted" yet about five milliard years for it, and this though we have not got time enough to follow these traces should be taken as an excuse in favor of us - they might in further future appear useful as antitemplates in bioregulations. At present our rather insufficiently fitted strands depending on their base and strand-polarity compositions exhibit effects in in vitro calf-thymus polymerase B (DNA-dependent RNA-polymerase) (71,135) and AMV-revertase (RNA-dependent DNA-polymerase) assays (70,142) (Figures 44 and 45), the stimulatory and inhibitory activities in the eukaryotic and the viral polymerase systems being intriguingly slightly counter-current with regard to the molecular strand design. The antitemplate effects, that seem to be mainly a reflection of the interhybrid recognitions rather than an expression of the probably less pronounced plastic-protein interactions, are completed by the assistant role of the polymer strands in inducing antiviral protection in mouse L-cells when base-paired with their mate polynucleotides. While partial $2(U)_n \cdot (A)_n$- and $2(I)_n \cdot (A)_n$-analogues only succeeded in establishing a slight antiviral status, the latter in addition demonstrating some priming effects (Figure 46) (78,143), partial $(I)_n \cdot (C)_n$-simulations like their famous all-polynucleotide ideal induce interferon, the interferon induction by the $(C)_n \cdot (vH,vCOONa)'_n$ representing the first example gained with a drastically altered $(I)_n$-component (Figures 36 and 47) (62).

These biological experiments confirm similar results of other groups (59,133) - mainly in the field of homopolymeric plastics - and illustrate the possibilities of our today polynucleotide-like mesogens to model and simulate even at their present stages of stereoelectronic insufficiencies rather complex biopolymer interaction. Thus in our eyes it seems a little bit encouraging, that not only the base-overcrowded homopolymers but also the more poorly base-fitted copolymers - which display on the other hand developing facilities of adjusting their variable

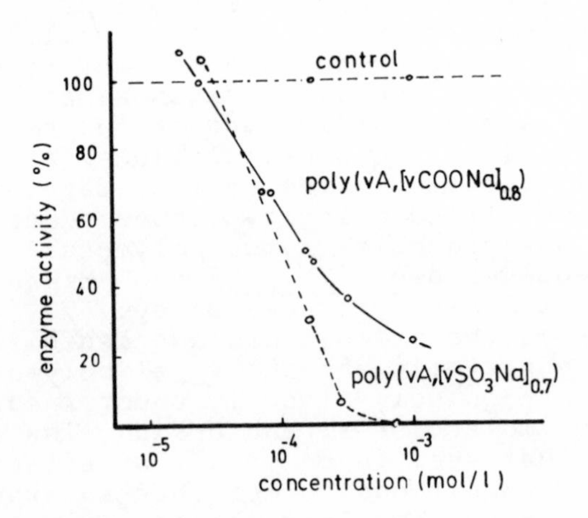

Zeitschrift fuer Chemie

Figure 44. Effects of some polynucleotide strand analogues on the transcription of CT-DNA by DNA-dependent RNA-polymerase (B) in vitro (71): Assay: pH 7.9; 0.1M Tris-HCl; 2 mM Mn^{2+}; 0.1 mM DTT; 0.02 mM BTPs; ^{32}P-ATP and -UTP; ^{3}H-ATP and -UTP; 8-10 γ DNA; T = 37°C. $(vA, [vOH]_{0.7})_n$; $(vA, [vCONH_2]_{1.1})_n$; $(vA, [vPn]_{0.8})_n$; $(vA, [vP]_{0.9})_n$; $(vA, [vI]_{1.1})_n$; $(vA, [vm^3I]_{0.6})_n$; $(vU, [vCOONa]_{1.4})_n$' like control without definite effect.

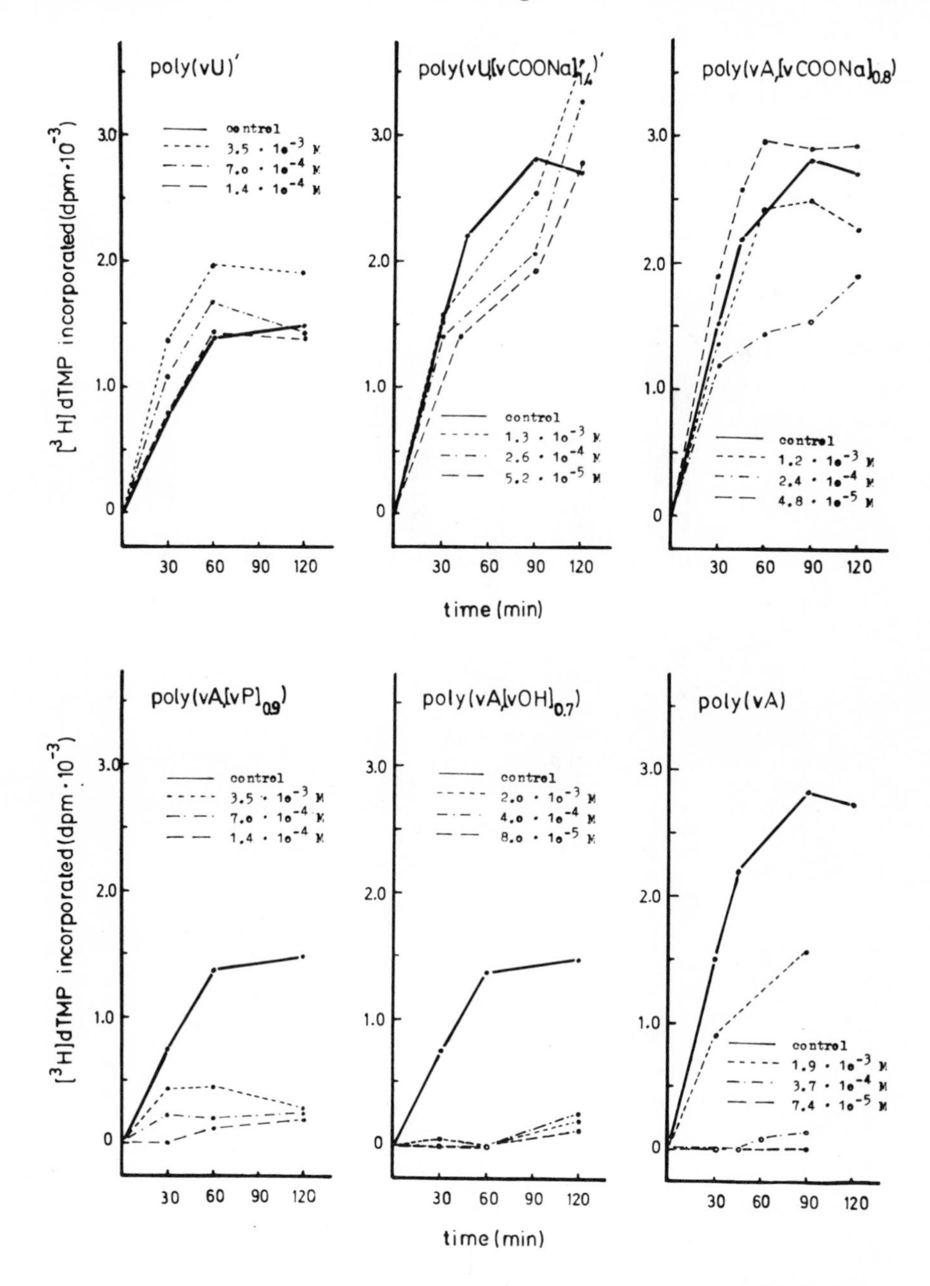

Zeitschrift fuer Chemie

Figure 45. Effects of some polynucleotide strand analogues in an in vitro AMV-revertase assay (RNA-dependent DNA polymerase; endogenous template-primer combination) (70). Assay: pH 8.0; 50 mM Tris-HCl; 2 mM DTT; 5 mM Mg^{2+}; 1 mM Mn^{2+}; 50 mM NaCl; 0.2 mM dBTPs; 10 μC ^3H-TTP; T = 37°C.

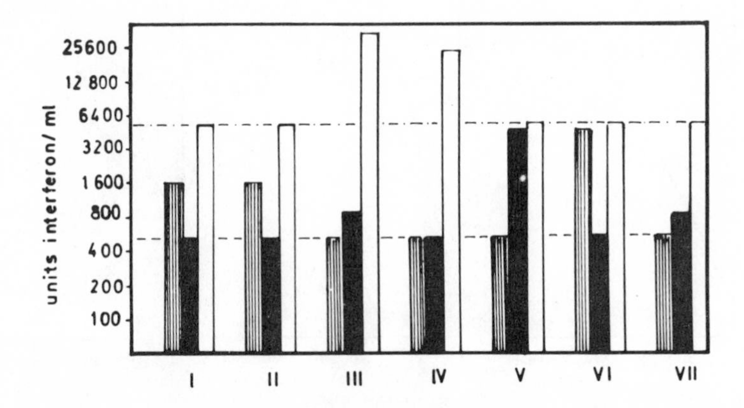

Figure 46. Effects of preincubation of mouse L-cells with polynucleotide triplexes and their partial analogues on subsequent interferon induction by NDV and on VSV propagation (78, 143). Assays: 10 μg/ml DEAE-dextran; T = 37°C; pH 7.2; PBS; concentration of the complexes 10^{-4}M (sum of bases). (a) Interferon priming: (– – –) control; (— · —) priming only with DEAE-dextran; (▤) 3 hr-treatment without DEAE-dextran; (■) 3 hr-treatment with DEAE-dextran; (▭) 24 hr-treatment with DEAE-dextran. (b) Antiviral states (c = VSV-test, c_o = VSV control).

	Complex	VSV-yield log PFU [c/c$_o$]
I	$(vA, [vOH]_{0.7})_n \cdot 2(U)_n$	−0.7
II	$(vA, [vCOONa]_{0.8})_n \cdot 2(U)_n$	−0.6
III	$(vA, [vOH]_{0.7})_n \cdot 2(I)_n$	−0.5
IV	$(vA, [vCOONa]_{0.8})_n \cdot 2(I)_n$	−1.0
V	$(vH,vCOONa) \cdot 2(A)_n$	−0.7
VI	$(A)_n \cdot 2(U)_n$	−0.8
VII	$(A)_n \cdot 2(I)_n$	−0.5

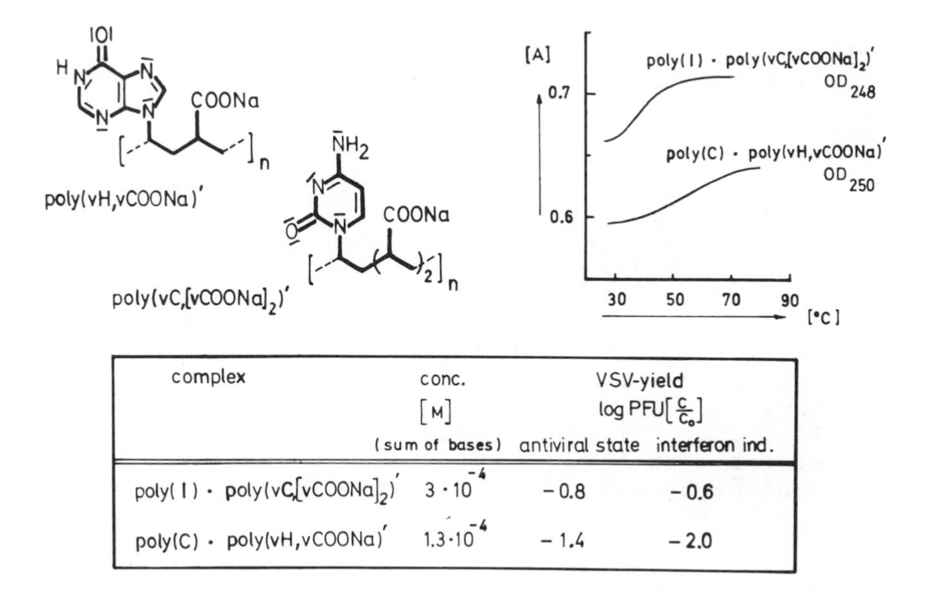

complex	conc. [M] (sum of bases)	VSV-yield log PFU[$\frac{c}{c_0}$] antiviral state	interferon ind.
poly(I) · poly(vC,[vCOONa]$_2$)'	$3 \cdot 10^{-4}$	−0.8	−0.6
poly(C) · poly(vH,vCOONa)'	$1.3 \cdot 10^{-4}$	−1.4	−2.0

Zeitschrift fuer Chemie

Figure 47. Establishment of antiviral states and interferon induction by partial $(I)_n \cdot (C)_n$-simulations in a mouse L-cell system; interferon induction by $(C)_n \cdot (I)_n$: −3.5 log PFU [c/c_0] for comparison. Other conditions of assaying like Figure 46 (62).

strand design at least to a certain degree to wanted levels of efficiency - reveal activites, which might be regarded (assisted by the results of the Mertes group (133) of blocking defined proteins) not only in terms of the mentioned obstruction of metabolic pathways by unspecific interactions, but perhaps as early beginnings of a biomesogen simulation chemistry, trying to reach the goals of native biopolymers and by this, developing their mesogen aspects from the undirectioned games of large ensembles of rather "unintelligent" individuals to the directioned domain cooperativities of "intelligent" stereoelectronic units.

Final Remarks

Yet at this point the question remains nevertheless without any definite answer, if there is any sense in modelling the highly evolved biopolymeric mesogens of our today life by dull and at this stage of refinement oversimplified analogues, with the presumption of using those simulations after a thinkable process of refinement and adjustment for assisting and even improving optimized pattern of natural efficiency.

It is part of the question wether man-made stereoelectronic pattern could act not only as competitors, but might advance for substitutes and partners of the natives of the great process.

Literature Cited

1. Neumann,E., Katchalsky,A., Ber.Bunsenges.Phys.Chem. (1970) 74, 868
2. Neumann,E.,Angew.Chem.(Int.Ed.)(1973) 12,356
3. Neumann,E.,in "25.Mosbacher Colloqu.Ges.Biol.Chem." 1974,(L.Jaenicke,ed.) Springer Verlag, Berlin Heidelberg, New York 1974
4. Spodheim,M.,Neumann,E.,Biophys.Chem.(1975) 3,109, and preceding communications
5. Cox,R.A.,Jones,A.S.,Marsh,G.E.,Peacocke,A.R.,Biochim.Biophys.Acta(1956) 21,576
6. Neumann,E.,Katchalsky,A.,Proc.Nat.Acad.Sci.USA (1972) 69,993
7. Hoffmann,S.,Witkowski,W.,Z.Chem.(1975) 15,149
8. Oster,G.,Perelson,A.,Katchalsky,A.,Nature(1971) 234,393
9. Sarvechi,M.-Th.,Guschlbauer,W.,Eur.J.Biochem.(1973) 34,232
10. Guschlbauer,W.,Thiele,D.,Sarvechi,M.-Th.,Marck,Ch., "Phénomènes d'hystérèse dans les polynucléotides"

in "Dynamic Aspects of Conformation Changes in Biological Macromolecules",(C.Sadron,ed.),Reidel-Publ. Co. Dordrecht-Holland 1973

11.Revzin,A.,Neumann,E.,Biophys.Chem.(1974) 2,144

12.Hoffmann,S.,Witkowski,W.,Rüttinger,H.-H.,"Hysteretic Metastabilities in the 2Poly(I)·Poly(A)-Triplex" Conf.Proc.IUPAC 4th Disc.Conf."Heterogenities in Polymers",1974,Marianske Lazne,Czechoslovakia

13.Hoffmann,S.,Witkowski,W.,Z.Chem.(1974) 14,438

14.Hoffmann,S.,Witkowski,W.,Rüttinger,H.-H.,"Molecular Hysteresis in Biopolyelectrolytes",Conf.Proc.1st Liquid Crystal Conf.of Socialist Countries,1976, Halle/S.,GDR

15.Mevarech,M.,Neumann,E.,Progr.Rep.Weizmann Inst., Rehovot Israel 1972

16.Mager,P.P.,Arzneimittel-Forsch.(1976) 26,1818

17.Mager,P.P.,"Chronobiologie",Leopoldina-Symp.,1975, Halle/S.,GDR

18.Tasaki,I.,Barry,W.,Carnay,L. in "Physical Principles of Biological Membranes" (F.Snell,ed.),Gordon-Breach,New York 1970,p.17

19.Clark,H.R.,Strickholm,A.,Nature(1971) 234,470

20.Träuble,H.,"Phase Transitions in Lipids" in "Biomembranes",Vol.3,p.197,(F.Kreuzer,J.F.G.Slegers,ed.) Plenum Publ.Corp.,New York

21.Träuble,H.,Eibl,H.,Proc.Nat.Acad.Sci.USA(1974) 71, 214, and preceding communications

22.Sackmann,E.,Ber.Bunsenges.(1974) 78,929

23.Nachmansohn,D.,Proc.Nat.Acad.Sci.USA(1976) 73,82, and preceding communications

24.Schneider,F.W.,Biopolymers(1976) 15,1

25.Bhat,R.K.,Schneider,F.W.,Ber.Bunsenges.Phys.Chem. (1976) 80,1153

26.Hoffmann,S.,Witkowski,W.,"Polynucleotide Strand-Analogues" and "Base-Pair Analogues" in Conf.Proc. "Wirkungsmechanismen von Herbiciden und synthetischen Wachstumsregulatoren" 1972,RGW-Symp.Halle/S., GDR

27.Langenbeck,W.,Halle/S. 1952-1966 personal communications

28.Arnott,S.,Bond,P.J.,Smith,P.J.C.,Nucleic Acids Res. (1976) 3,2459

29.Torrence,P.F.,DeClercq,E.,Witkop,B.,Biochim.Biophys.Acta(1977) 475,1, and preceding communications

30.DeClercq,E.,Topics in Current Chemistry (1974) 52, 174

31.Gillespie,D.,Gallo,R.C.,Science (1975) 188,802

32.Carter,W.A.,DeClercq,E.,Science (1974) 186,1172

33.Demus,D.,Demus,M.,Zaschke,H.,"Flüssige Kristalle in Tabellen",VEB Deutscher Verlag für Grundstoffindustrie,Leipzig,1974

34. Schubert,H.,Wiss.Z.Univ.Halle (1970) 19,1
35. Schubert,H., Halle/S. 1957 and following years: personal communications
36. Clough,S.B.,Blumstein,A.,Hsu,E.C.,Macromolecules (1976) 9,123
37. Blumstein,A.,Weill,G.,Macromolecules (1977) 10,75, and preceding communications
38. Perplies,E.,Ringsdorf,H.,Wendorf,J.H.,Ber.Bunsenges.Phys.Chem. (1974) 78,921
39. Wendorff,J.H.,Perplies,E.,Ringsdorf,H.,Progr.Colloid & Polymer Sci. (1975) 57,272, and preceding communications
40. Cser,T.,Nyitrai,K.,Ngoc,B.D.,Seyfried,E.,Hardy,G., "Investigations on Polymer-Containing Liquid Crystals",Conf.Proc.1st Liquid Crystal Conf. of Socialist Countries,1976,Halle/S.,and preceding communications
41. Hardy,G.,Nyitrai,K.,Cser,T.,"Polymerization and Copolymerization in Mesomorphic Phases",Conf.Proc.1st Liquid Crystal Conf. of Socialist Countris,1976, Halle/S.,GDR, and preceding communications
42. Baturin,A.A.,Amerik,Y.B.,Krentsel,B.A.,Mol.Cryst. Liquid Cryst.(1972) 16,117, and preceding communications
43. Strzelecki,L.,Liebert,L.,Keller,P.,Bull.Soc.Chim. France (1975) 2750, and preceding communications
44. Tanaka,Y.,Hitotsuyanagi,M.,Shimura,Y.,Okada,A., Sakuraba,H.,Sakata,T.,Makromol.Chem. (1976) 177, 3035
45. Kamogawa,H.,Polymer Letters (1972) 10,7
46. Conf.Proc.IUPAC 5th Disc.Conf.PMM "Phases and Interfaces in Macromolecular Systems", 1976,Praha, Czechoslovakia
47. Fernandez-Bermudez,S.,Balta-Calleja,F.J.,Hosemann, R.,Makromol.Chem. (1974) 175,3567, and preceding communications
48. Staab,H.A.,Angew.Chem. (1962) 74,407
49. Hoffmann,S.,Witkowski,W.,Weißpflog,W.,Borrmann,G., manuscript in preparation
50. Hoffmann,S.,Brandt,W.,Schubert,H.,Z.Chem. (1975) 15,59
51. Hoffmann,S.,Brandt,W.,Z.Chem. (1975) 15,306
52. Hoffmann,S.,Brandt,W.,Schubert,H.,"Mesogen Steroid-Hormone Derivatives",Conf.Proc.1st Liquid Crystal Conf. of Socialist Countries,1976,Halle/S.,GDR
53. Hoffmann,S.,Brandt,W., unpublished results
54. Hoffmann,S.,Weißpflog,W.,Kumpf,W.,Witkowski,W., Brandt,W.,manuscript in preparation
55. Hoffmann,S.,Witkowski,W.,Borrmann,G.,manuscript

in preparation
56. Hoffmann,S.,Witkowski,W.,Borrmann,G.,Z.Chem. (1977) 17,291
57. Meindl,P.,Bodo,G.,Tuppy,H.,Arzneimittel-Forsch. (1976) 26,312
58. Kaye,H.,Chang,S.-H.,Macromol.Sci.-Chem. (1973) A 7,1127, and preceding communications
59. Pitha,J.,Polymer,(1977) 18,425 and preceding communications
60. Inaki,Y.,Takada,H.,Kondo,K.,Takemoto,K.,Makromol. Chem. (1977) 178,365
61. Seita,T.,Yamauchi,K.,Kinoshita,M.,Imoto,M.,Makromol.Chem. (1973) 164,7, and preceding communications
62. Hoffmann,S.,Witkowski,W.,Waschke,K.,Waschke,S.-R., Z.Chem. (1977) 17,61, and preceding communications
63. Hoffmann,S.,Witkowski,W.,Gyulbudagyan,A.,Z.Chem. (1977) 17,102
64. Hoffmann,S.,Witkowski,W.,Nucleic Acids Res. (1975) S 1,s137
65. Hoffmann,S.,Witkowski,W.,Ladhoff,A.-M.,Z.Chem. (1976) 16,228
66. Hoffmann,S.,Witkowski,W.,Behlke,J.,Z.Chem. (1976) 16,275
67. Hoffmann,S.,Witkowski,W.,Schubert,H.,Salewski,D., Kölling,M.-L.,Z.Chem. (1974) 14,309
68. Hoffmann,S.,Witkowski,W.,Frahnert,Ch.,Mehnert,Ch., unpublished results
69. Hoffmann,S.,Witkowski,W.,Mehnert,Ch.,Geserick,G., unpublished results
70. Hoffmann,S.,Witkowski,W.,Venker,P.,Rößler,H., Z.Chem. (1976) 16,404
71. Hoffmann,S.,Witkowski,W.,Grade,K.,Schönheit,Ch., Z.Chem. (1976) 16,324
72. Hoffmann,S.,Witkowski,W.,Rüttinger,H.-H., manuscript in preparation
73. Felsenfeld,G.,Miles,H.T.,Ann.Rev.Biochem. (1967) 36,407
74. Michelson,A.M.,Massoulié,Guschlbauer,W.,Progr. Nucl.Acid Res.Mol.Biol. (1967) 6,83
75. Hoffmann,S.,Witkowski,W.,Munsche,D., manuscript in preparation
76. Hoffmann,S.,Witkowski,W.,Luck,G.,Zimmer,Ch., manuscript in preparation
77. Hoffmann,S.,Witkowski,W.,Luck,G.,Zimmer,Ch.,Skölziger,R.,Veckenstedt,A.,manuscript in preparation
78. Hoffmann,S.,Witkowski,W.,Waschke,K.,Z.Chem. (1976) 16,484
79. Hoffmann,S.,Witkowski,W.,Veckenstedt,A., unpublished results

80.Watson,J.D.,Crick,F.H.C.,Nature (1953) 177,964
81.Dickerson,R.E.,Takano,T.,Eisenberg,D.,Kallai,O.B., Samson,L.,Cooper,A.,Margoliash,E.,J.Biol.Chem. (1971) 246,1511
82.Thiele,D.,Guschlbauer,W.,Biophysik (1973) 9,261
83.Arnott,S.,Chandrasekaran,R.,Marttila,C.M.,Biochem. J. (1974) 141,537
84.Zimmerman,S.B.,Cohen,G.H.,Davies,D.R.,J.Mol.Biol. (1975) 92,181
85.Zimmerman,S.B.,J.Mol.Biol. (1976) 106,663
86.Rich,A.,RajBhandary,U.L.,Ann.Rev.Biochem. (1976) 45,806
87.Ladner,J.E.,Jack,A.,Robertus,J.D.,Brown,R.S., Rhodes,D.,Clark,B.F.C.,Klug,A.,Proc.Nat.Acad.Sci. USA(1975) 72,4414, and preceding communications
88.Saenger,W.,Riecke,J.,Suck,D.,J.Mol.Biol. (1975) 93,529
89.Arnott,S.,Chandrasekaran,R.,Leslie,A.G.W.,J.Mol. Biol. (1976) 106,735
90.McGavin,S.,J.Mol.Biol. (1971) 55,293
91.Lindigkeit,R.,Böttger,M.,v.Mickwitz,C.-U.,Fenske, H.,Karawajew,L.,Karawajew,K.,Acta biol.med.germ. (1977) 36,275
92.Dina,D.,Meza,I.,Crippa,M.,Nature (1974) 248,486
93.Arnott,S.,Bond,P.J.,Nature (1973) 244,99, and Science (1973) 181,68
94.DeClercq,E.,Torrence,P.F.,DeSomer,P.,Witkop,B., J.Biol.Chem. (1975) 250,2521
95.Hoffmann,S.,Witkowski,W.,Z.Chem. (1976) 16,442
96.Rosenberg,J.M.,Seeman,N.C.,Day,R.O.,Rich,A., Biochim.Biophys.Res.Commun. (1976) 69,979
97.Hoffmann,S., Witkowski,W., "Polynucleotide Strand-Analogues" and "Base-Pair Analogues" in "Wirkungsmechanismen von Herbiciden und synthetischen Wachstumsregulatoren" (A.Barth,F.Jacob,G.Feyerabend,eds.) VEB Gustav Fischer Verlag, Jena 1975
98.Hoffmann,S.,Witkowski,W.,"Negativmodellierungen von Rezeptorregionen", Conf.Proc."Problems in Bioeffector Research" 1977,Oberhof,GDR and Wiss.Z. Univ.Halle in the press
99.Hoffmann,S.,Witkowski,W.,Schubert,H.,Z.Chem. (1974) 14,154
100.Hoffmann,S.,Witkowski,W., Conf.Proc.RGW-Symp. "Wirkungsmechanismen von Herbiciden und synthetischen Wachstumsregulatoren", 1972, Halle/S.,GDR
101.Jensen,E.V.,DeSombre,E.R.,Science (1973) 182,126
102.Duax,W.L.,Wecks,C.M.,Rohrer,D.C.,"Crystal Structures of Steroids" in "Stereochemistry" (Allinger and Eliel,eds.) John Wiley & Sons Inc. 1976
103.Pohlmann,J.L.W.,Elser,W.,Boyd,P.R.,Mol.Cryst.Li-

quid Cryst. (1971) 13,225, and preceding communications
104.Malthet,J.,Billard,J.,Jacques,J.,Bull.Soc.Chim. France (1974) 1199
105.Gabbay,E.J.,Glaser,R.,Biochemistry (1971) 10,1665
106.Waring,M.J.,Chisholm,J.W.,Biochim.Biophys.Acta (1972) 262,18
107.Vorländer,D.,Janecke,F.,Z.Phys.Chem. (1913) A85, 697
108.Johnson,W.C.,Tinoco,I.,Biopolymers (1969) 7,727
109.Stegemeyer,H., Halle/S. 1977, personal communication
110.Stegemeyer,H.,Mainusch,K.-J.,Naturwissenschaften (1971) 58,599
111.Stegemeyer,H.,Ber.Bunsenges.Phys.Chem. (1974) 78, 861
112.Waring,M.,Chem.&Ind.,(1975) 105
113.Pigram,W.J.,Fuller,W.,Davies,M.E.,J.Mol.Biol. (1973) 80,361
114.Pigram,W.J.,Fuller,W.,Hamilton,L.D.,Nature (1972) 235,17
115.Trai,Ch.-Ch.,Jain,S.C.,Sobell,H.M.,Proc.Nat.Acad. Sci.USA (1975) 72,628, and preceding communications
116.Carr,A.A.,Grunwell,J.F.,Sill,A.D.,Meyer,D.R., Sweet,F.W.,Scheve,B.J.,Grisar,M.,Fleming,R.W., Mayer,G.D.,J.Med.Chem. (1976) 19,1142,and preceding communications
117.Mayer,G.D.,Interscience Conf."Antimicrobial Agents and Chemotherapy",1974,San Francisco,USA
118.Chandra,P.,Woltersdorf,M.,Biochem.Pharmacol. (1976) 25,877
119.Daune,M.,Sturm,J.,Zana,R.,stud.biophys. (1976) 57,139
120.Meindl.P.,Bodo,G.,Tuppy,H.,Drug-Res. (1976) 26,312
121."Analogs of Nucleosides and Nucleotides and Their Applications",19.Conf.Ges.Biol.Chem. Hoppe-Seyler's. Z.Physiol.Chem. (1974) 355,753
122.Prusoff,W.H.,Ward,D.C.,Biochem.Pharmacol. (1976) 25,1233
123.Shugar,D.,FEBS Letters (1974) 40,548
124.Jones,A.S.,Walker,R.T.,Nucleic Acids Res. (1975) S1,s109, and preceding communications
125.Hoffmann,S.,Schubert,H.,Witkowski,W.,Z.Chem. (1971) 11,345
126.Jones,A.S.,Markham,A.F.,Walker,R.T.,Tetrahedron (1976) 32,2361, and preceding communications
127.Thiellier,H.P.M.,Koomen,G.J.,Pandit,U.K.,Tetrahedron (1977) 33,1493, and preceding communications
128.Shomshtein,Z.A.,Hiller,S.A.,Chem.Heterocycl.Comp.

(USSR)(1976) (1),27, and preceding communications
129.Kinoshita,M.,Yamauchi,K.,Imoto,M.,"Synthetic Polymers containing Nucleic Acid Bases and Their Derivatives" preprint 1974, and preceding communications
130.Pitha,J.,Pitha,P.M.,Stuart,E.,Biochemistry (1971) 10,4595, and preceding communications
131.Holý,A.,Coll.Czechoslov.Chem.Commun. (1975) 40, 187, and preceding communications
132.Walker,R.T.,Ann.Rep.Chem.Soc. (1973) 3,624
133.Boguslawski,S.,Olson,P.E.,Mertes,M.P.,Biochemistry (1976) 15,3536, and preceding communications
134.Mohr,S.J.,Brown,D.G.,Coffey,D.S.,Nature (1972) 240,250
135.Hoffmann,S.,Schubert,H.,Witkowski,W.,Z.Chem. (1971) 11,465
136.Bugg,C.E.,Thomas,J.M.,Sundaralingam,M.,Rao,S.T., Biopolymers (1971) 10,175
137.Bugg,C.E.,"The Purines - Theory and Experiment" in "Jerusalem Symp. on Quant.Chem. and Biochem.IV" Israel Academy of Sciences and Humanities Jerusalem 1972
138.Mislow,K.,"Introduction to Stereochemistry", W.A. Benjamin Inc., 1965
139.Kusanagi,H.,Tadokoro,H.,Chatani,Y.,Macromolecules (1976) 9,531
140.Motherwell,W.D.S.,Isaacs,N.W.,J.Mol.Biol. (1972) 71,231
141.Thrierr,J.C.,Dourlent,M.,Leng,M.,J.Mol.Biol. (1971) 58,815
142.Hoffmann,S.,Witkowski,W.,Venker,P.,Grade,K.,unpublished results
143.Waschke,K.,Waschke,S.-R.,Hoffmann,S.,Witkowski,W., "Interferon Induction by Double-Strandes Like Complexes of Vinyl Copolymers with Polynucleotides" in Conf.Proc. 4th Regional Symp. "Interferon and Interferon Inducers" Wroclaw, Poland, October 1976 in the press

RECEIVED December 8, 1977.

15

Cholesteric Order in Biopolymers

Y. BOULIGAND

Histophysique et Cytophysique, E.P.H.E. et C.N.R.S., 67, rue Maurice-Günsbourg, 94200 Ivry-sur-Seine, France

Many synthetic polymers form cholesteric phases, and even solids showing certain of the fundamental symmetries of cholesteric liquids. The purpose of this paper is to review the main examples of biological polymers assembling into cholesteric liquids or into more or less solid analogues. We will present them according to the main chemical classes of polymers to which they belong. We will also indicate the main forces involved in creating the cholesteric twist.

Analogues of liquid crystals are numerous in biological systems (3-6,8-11). Certain fibrous and regularly twisted materials can be considered as polymerized cholesterics. Such twisted fibrous structures are recognizable by electron microscopy and sometimes by light microscopy, by the observation in thin sections of sets of stacked rows of nested arcs (fig. 1). The structure of a cholesteric system is represented in fig. 2 and the origin of the arced patterns is indicated in fig. 3. The first sketch (fig. 2) shows a section which is parallel to the twist axis. Fig. 3 corresponds to a section plane which is oblique with respect to the twist axis. This is the case most frequently observed.

Pictures of stacked series of nested arcs in biological analogues of cholesterics have been found in numerous invertebrate materials. They have been reviewed in (6). New references can be found in (20) and (21). Twisted fibrous arrangements have been observed in the organic matrix of compact bone in vertebrates (6,23 and Castanet pers. comm.), the connective tissue of certain invertebrates (6,14 and fig. 1), in numerous animal cuticles (3-6,8-10,20,30), in the body-wall of certain Tunicates (6,15), in the membranes surrounding various animal eggs (6), in various types of cytoplasmic inclusions (6), in different kinds of plant cell walls (6,21,22,27), in certain plant mucilages (J.-C.Thomas, pers. comm.), in the bacterial nucleus (6, Gourret, pers. comm.) and in the chromosomes of certain Protozoa (9, figs 6 and 7).

Nucleic acids, proteins and glycoproteins are the main components of the twisted fibrous arrangements. For many of them, biochemical studies are rare or absent.

0-8412-0419-5/78/47-074-237$05.00/0

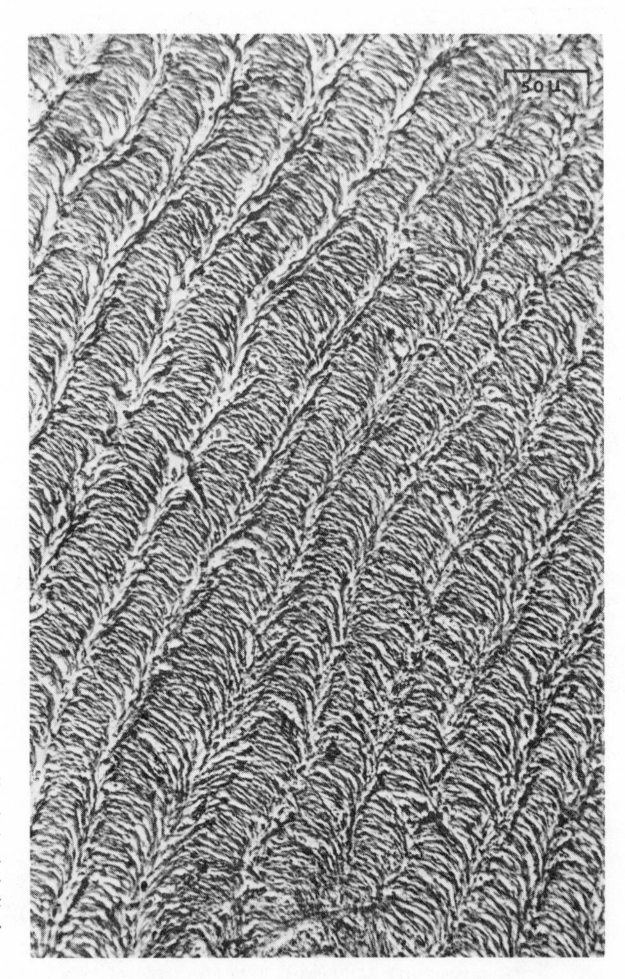

Figure 1. Stacked series of nested arcs observed in oblique section in the connective tissue of Havelockia inermis, *Holothuroid, Echinoderm. Phase contrast microscopy, paraffin section, Haematoxylin.*

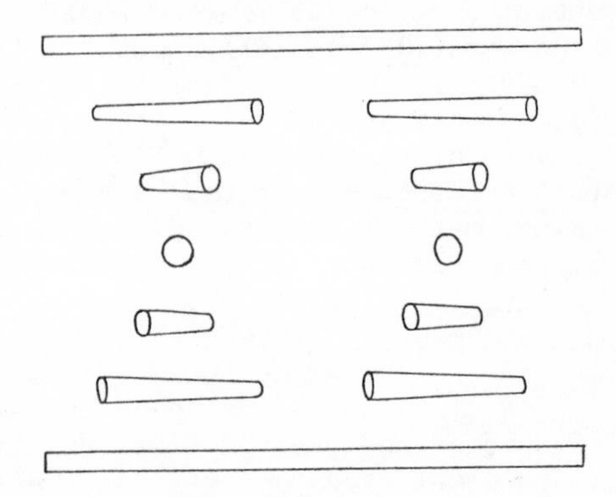

Figure 2. Cholesteric architecture observed in a section plane parallel to the twist axis

1. Nucleic acids. The DNA of the Dinoflagellate chromosomes forms a cholesteric network which is very similar to that often observed in the bacterial nucleus (6,9,18). Similar structures have been observed in mitochondrial DNA in certain Trypanosomes, after a treatment with certain drugs (review in 6). DNA in concentrated aqueous solutions can form cholesteric mesophases (4,18, 24). Remarkable cholesteric spherulites have been observed in concentrated ribosomal RNA (first considered as t-RNA, 33-35).

2. Proteins. Synthetic polypeptides can form cholesteric solutions in several organic solvents (24-26,29). Twisted arrangements of collagen fibrils are common in sponges (Carrière, pers. comm.), in Holothurians (Echinoderms, 6,14, see fig. 1) and in flat worms (6).

Microtubular haemoglobin has been observed in erythrocytes in sickle cell anaemia. After deoxygenation and addition of thermal energy, the mutant molecules of haemoglobin (HbS) stack to form monomolecular filaments. Six strands assemble into a helicoidal microtubule showing six helices of long pitch (review in 19). These microtubules form a cholesteric packing (6).

Superb twisted arrangements have been observed in larval haemocytes (oenocytoid) of the silkworm, forming regular and concentric series of arcs (2).

Proteins also are known to form a twisted fibrous system in the periostracum of certain gastropods (review in 20) and in the cortex of the oocytes of numerous Teleost fishes, review in (6).

3. Glycoproteins (associated polysaccharides and proteins). Cellulose and proteins form microfibrils in the body-wall of certain marine invertebrates, the Tunicates. The fibrous network shows a regular twist in one species *Halocynthia papillosa*. Cellulose is an important component of plant cell walls and series of nested arcs are often visible. Interesting pictures have been discussed with reference to the cholesteric conception (21,22,27). Plant cell walls contain a great variety of polysaccharides and they are often different from cellulose. The examples indicating the presence of a twisted architecture are reviewed in (6,21,22, 27). Certain blue-green algae (*Rivularia atra* and *Chroococcus minutus*) show remarkable arced patterns in their mucilage (J.-C. Thomas, pers. comm. fig. 8).

Chitin/Protein complexes are found in fungi and in the arthropod cuticle. Chitin is the poly-N-acetyl-D-glucosamine. Twisted arrangements have been observed in the cell wall of the spore of *Endogone* (Mucorale, fungi) and in various animal cuticles, namely in certain medusae (6) and in almost all Arthropods: Crustaceans (3-6), Insects and other groups (review in 20).

4. Twisted systems due to viruses. Certain viruses in the form of long cylinders can assemble into cholesteric phases (Narcissus mosaic virus, 6,39). The presence of a virus in a cell often leads to the differentiation of cholesteric networks of fibrils (6,16).

Many works deal with the variation of the helicoidal pitch in cholesteric phases as a function of temperature and composition. The best way to elucidate the origin of the twist seems to be to compare the pitch variations in lyotropic and in thermotropic systems. The first accurate work in this field is due to Robinson (1958-66) who studied PBLG (polybenzyl-L-glutamate) a synthetic polypeptide in organic solvents as dioxane, ethylic alcohol, chloroform etc. and Cano (1967) who made measurements of the pitches of nematic paraazoxyphenetol with different amounts of cholesterol benzoate.

The twist is proportional to the concentration of the twisting molecules in Cano's experiments and to the square of the concentration in Robinson's. These observations suggest that this twist is proportional to the frequency of molecular collisions between cholesteric and nematic molecules (C and N) in Cano's work and between two cholesteric molecules (C with C) in Robinson's. Accordingly, it appears that the twist can be derived from the molecular concentrations in an analogous way to that of the chemical kinetics (Bouligand, 1974).

More generally, in a cholesteric phase, different kinds of molecules may be involved and belong to four main types:

N: molecules able to form by themselves a nematic phase.
C: molecules able to form by themselves a cholesteric phase.
R: non-mesogenic molecules showing a molecular rotatory power and able to twist a nematic liquid.
S: non-mesogenic solvent without any twisting influence.

The collisions between two molecules susceptible to create an elementary twist are of the following types

N+C; N+R; C+C; C+R.

The resulting twist is a linear expression of the frequencies of the various kinds of molecular contacts and will be:

$$t_o = m_1[N][C]+m_2[N][R]+m_3[C][R]+m_4[C]^2 \quad (7),$$

where m_1, m_2, m_3, m_4 are constants at a given temperature and the brackets indicate concentrations. These formulations allow the interpretation of more recent results (1).

Cano's experiments correspond to

$$t_o = m_1[C][N]+m_4[C]^2, \quad \text{with } [C]+[N] = 1.$$

One has thus:

$$t_o = m_1[C]+(m_4-m_1)[C]^2$$

and for the range of concentrations studied by Cano $0<[C]<0.2$, one has a good proportionality between t_o and [C].

Robinson's experiments correspond to

$$t_o = m_4[C]^2.$$

Another view of the cholesteric twist is interesting to discuss. Helical configurations are common in the polymers forming cholesteric systems. Rudall (1955) has proposed a cholesteric model allowing a very tight packing of helical polymers. Consider for instance a set of double helices arranged in stacked layers. In one layer, helices are parallel, each being shifted by p/4 with respect to the next. This arrangement forms parallel and equidistant grooves, which lie obliquely with respect to the helical axes (fig. 4). A second layer of double helices can be arranged in these parallel grooves (Fig. 5). There may be a difference between the distance d of two successive helices in one layer and the distance e separating the grooves (fig. 4). So the helices of the second layer are not necessarily in exact register with the furrows of the first layer. The two systems may be out of phase. This vernier effect can slightly disturb this kind of packing resulting in a helicoidal plywood with discrete steps of rotation. One must note that right-handed helices lead to a left-handed cholesteric packing and conversely.

A similar model can be considered even when a system of polymers forms microfibrils or filaments distributed with a constant interdistance. The separation of filaments can be controlled by the balance of several forces: hydration, electrical double-layer repulsion and van der Walls attraction. This conception derives from a paper by Elliot and Rome (1969) who interpreted in that way the interdistance control between muscle cell filaments. We do not know the exact shape of the isopotential surfaces corresponding to that system of forces in the vicinity of fibrils formed by a set of chiral polymers, but they must show in many cases a helicoidal symmetry. The complete system can closely resemble the model suggested by Rudall (1955).

One must note that Rudall's model and its extension to systems of filaments showing a constant interdistance are compatible with the statistical conception elaborated from thermotropic and lyotropic cholesterics examined above. When polymers like P.B.L.G. form cholesteric solutions, one often observes an hexagonal packing (26). This means that a polymer can hesitate between the furrow separating two parallel polymers and the oblique grooves due to the helical structure. More generally, molecules and polymers are submitted to the molecular field which tends to align them parallel to a given direction and forces due to the chirality of molecules. These two behaviours occur simultaneously and their ratio determines the helicoidal pitch. Our first mentioned statistical conceptions with collisions correspond more to liquid systems and the generalized model of Rudall is better adapted to more or less consolidated gels. When α-helices of PBLG are arranged parallel to form one layer, it has been shown how the helices due to benzyl groups form oblique equidistant grooves (36) and this example appears to be a good illustration of Rudall's model. In its generalized form, this model seems to suit perfectly some micrographs of both Crustacean (3) and Insect

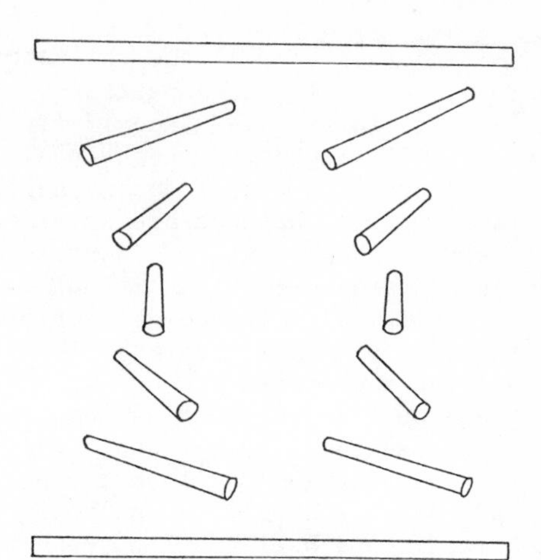

Figure 3. Oblique section of a cholesteric structure. The molecules appear to be aligned along arcs in the plane of the section.

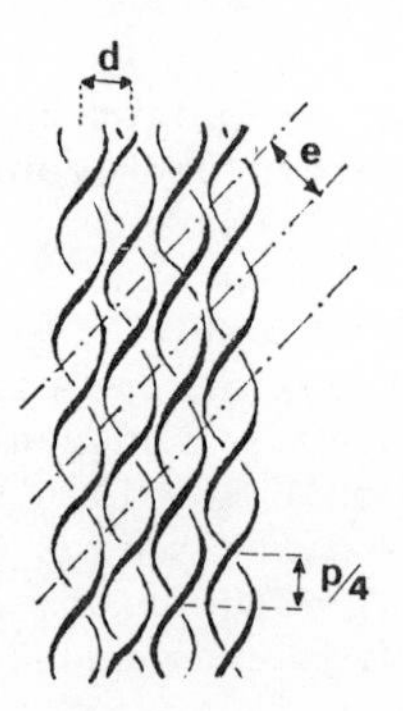

Figure 4. Double helices forming a dense layer; d *distance separating the helical axes,* e *distance of the grooves,* p *helical pitch*

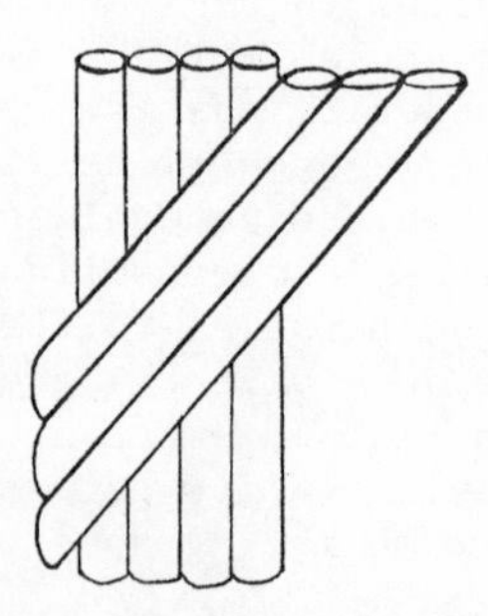

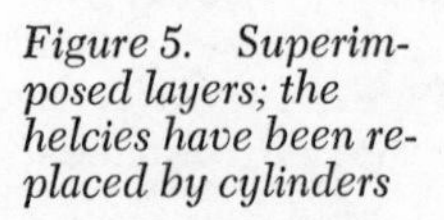

Figure 5. Superimposed layers; the helcies have been replaced by cylinders

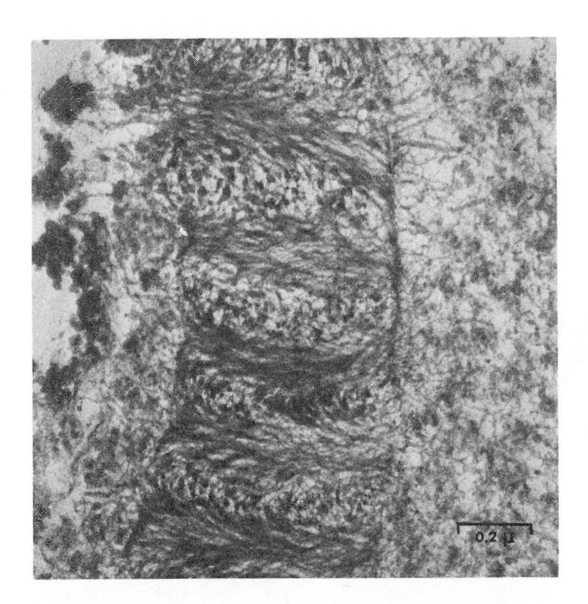

*Figure 6. Dinoflagellate chromosome (*Prorocentrum micans*), section plane parallel to the twist axis (courtesy of F. Livolant.)*

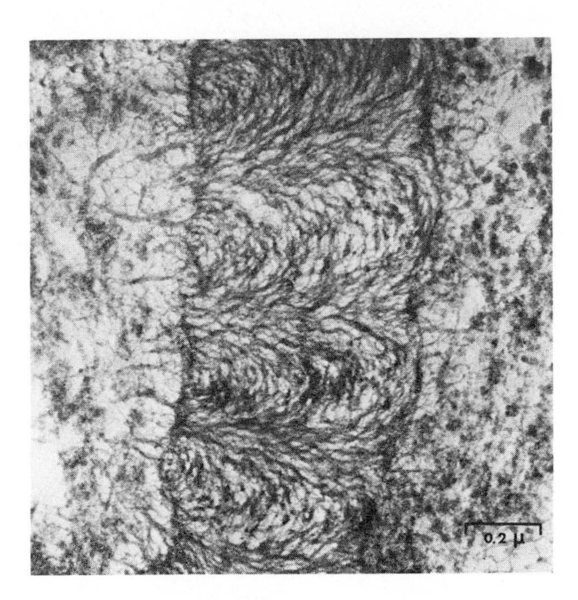

*Figure 7. Dinoflagellate chromosome (*Prorocentrum micans*), oblique sections. Arrows indicate the coiled bundles. (courtesy of F. Livolant).*

Figure 8. Arced patterns of the mucilage of Rivularia atra *(courtesy of J.-C. Thomas)*

(20) cuticles. Parallel alignments of supramolecular rods often lead to hexagonal patterns and these have to be limited to a small number of layers in order to allow the twist to occur.

In many instances, the helical pitch of fibrous biological materials is variable and, in Arthropod cuticle, the twist is controlled by the epidermis. During the cuticle secretion, the helical pitch varies according to a program repeated for each new cuticle. Renewing of the cuticle is imposed in Arthropods by discontinuous growth and periodic molts in which the old cuticle is shed. The variable pitch is not easily interpreted in the Rudall conception. It has been suggested (8) that a twisting factor could act exactly as a soluble chiral compound transforming certain nematic liquids into cholesteric liquids and variations in this twisting factor could account for variations in pitch. One finds steroids in the Arthropod cuticle but their role in the twist has not yet been demonstrated.

In chromosomes, certain micrographs (figs. 6 and 7) suggest the existence of coiled bundles of DNA filaments. The whole system is a twisted network of helicoidal bundles of helical polymers. These bundles are likely to be microcrystals of DNA filaments, either due to preparation of the material for electron microscopy, or to the simple fact that DNA is very concentrated in chromosomes. Indeed, the presence of proteins (17) and ribonucleic acids (12,31,32) facilitate the condensation of DNA. In such microcrystalline bundles, there are two possible arrangements: the hexagonal parallel packing or the dense twisted plywood of Rudall. Such coiled bundles have been observed electron microscopically in thin films formed by a dried PBLG solution. The right-handed double-helices of DNA lead to a left-handed mutual twist of the filaments and, therefore, a right-handed torsion for the bundles. The chirality of these bundles and that of double-helices cooperate to give a left-handed twist of the fibrous cholesteric arrangement of the whole chromosome. The left-handedness of the mutual twist of DNA in chromosomes has been demonstrated by stereo electron microscopy of dinoflagellate chromosomes (9).

A helicoidal model has been proposed for cellulose by Viswanathan and Shenonda (1971). This assumption needs further investigations and is interesting since there are many twisted systems formed by cellulose or related polymers. An example of arcs given by polysaccharidic microfibrils is shown in fig. 8.

We have also shown (6) that nematic analogues are found in many biological materials which show a strong birefringence. A great proportion of polymers forming condensed phases align to form birefringent systems. They have the fundamental symmetries of cholesteric and nematic liquid crystals. They are more or less hardened gels, this character being due to microcrystallizations. The material which is secreted earliest is either a true liquid crystal (birefringent and fluid), or a loose birefringent gel (Bouligand, 1975).

Literature Cited

1. Adams J. and Haas W., Molecular Crystals and Liquid Crystals (1975) 30, 1.
2. Akai H. and Sato S., Int. J. Insect Morph. Embryol. (1973) 2, 207.
3. Bouligand Y., C. R. Acad. Sci. Paris (1965) 261, 3665, 4864.
4. Bouligand Y., J. Physique (1969) 30(C4), 90.
5. Bouligand Y., J. Microscopie (1971) 11, 441.
6. Bouligand Y., Tissue and Cell (1972) 4, 189.
7. Bouligand Y., J. de Physique (1974) 35, 215.
8. Bouligand Y., J. Physique (1975) 36 (C1), 331.
9. Bouligand Y., Soyer M.-O. and Puiseux-Dao S., Chromosoma (1968) 24, 251.
10. Cano R., Bull. Soc. Fr. Minéral. (1967) 90, 333.
11. Chapman D., Ann. N.-Y. Acad. Sci. (1968) 137, 745.
12. Delius H. and Worcel A., Cold Spr. Harb. Symp. Quant. Biol. (1973) 38, 53.
13. Elliot G. F. and Rome E. M., Molecular Crystals and Liquid Crystals (1969) 8, 215.
14. Gross J. and Piez K. A.,"Calcification in Biological Systems" (1960), ed. by Sognnaes, A.A.A.S., 64.
15. Gubb D. C., Tissue and Cell (1975) 7, 1.
16. Horne R. W., Symp. Soc. Exptl. Biol. (1971) 25, 71.
17. Kowallik K. W., Arch. Mikrobiol. (1971) 80, 154.
18. Lerman L. S., Cold Spr. Harb. Symp. Quant. Biol. (1973) 38, 59.
19. Murayama M., Critical Rev. in Biochem. (1973) 1, 461.
20. Neville A. C.,"Biology of the Arthrop Cuticle" (1975) Springer.
21. Neville A. C., Gubb D. C. and Crawford R. M., Protoplasma (1976) 90, 307.
22. Peng H. B. and Jaffe L. F., Planta (1976) 133, 57.
23. Pritchard J. J., in "The Biochemistry and Physiology of Bone," G. H. Bourne ed. Academic Press (Plate 1, fig 7, p. 25), (1956).
24. Robinson C., Tetrahedron (1961) 13, 219.
25. Robinson C., Mol. Crystals Liq. Cr. (1966) 1, 467.
26. Robinson C., Ward J. C. and Beevers R. B., Far. Soc. Disc. (1958) 25, 29.
27. Roland J. C., Vian B. and Reis D., Protoplasma (1977) 91, 9.
28. Rudall, K. M., Lectures on the Scient. Basis of Medic. (1955) 5.
29. Samulski E. T. and Tobolski A. V. in"Liq. Crystals Plast. Crystals," ed. Gray and Winsor, Ellis Horwood (1974) 1, 175.
30. Shepherd A. M., Clark S. A. and Dart J., Nematologica (1972) 18, 1.
31. Soyer M.-O., Chromosoma (1971) 33, 70.
32. Soyer M.-O. and Haapala O. K., Chromosoma (1974) 47, 179.
33. Spencer M., Cold Spr. Harb. Symp. Quant. Biol. (1963) 28, 77.

34. Spencer M., Fuller W., Wilkins M. H. and Brown G. L., Nature (1962) 194, 1014.
35. Spencer M. and Poole F., J. Mol. Biol. (1965) 11, 314.
36. Squire J. M. and Elliot A., Molecular Crystals and Liquid Crystals (1969) 7, 457.
37. Viswanathan A. and Shenonda S. G., J. appl. Polym. Sci. (1971) 15, 519.
38. Vos L. de, J. Microscopie (1972) 15, 247.
39. Wilson H. R. and Tollin P., J. Ultrastructure Res. (1970) 33, 550.

RECEIVED December 8, 1977.

16

Liquid Crystalline Contractile Apparatus in Striated Muscle

ERNEST W. APRIL

Department of Anatomy, College of Physicians & Surgeons of Columbia University, New York, NY 10032

The ubiquity of liquid crystalline structures in living cells and the essential role of liquid crystrals in life processes has become evident. The lattice of myosin filaments, comprising the A or anisotropic band in striated muscle, can be likened to a smectic "B" liquid crystal. The I or isotropic band, composed of thinner actin filaments, does not usually form regular crystalline arrays in the muscle fiber. However, when the two populations of filaments interdigitate, as they must for the generation of tension through the interaction between actin and myosin, they form a double interleaving smectic array. The stability of the A band lattice in striated muscle appears to be due to a balance between van der Waal's attractive forces and long range electrostatic repulsive forces (1). These forces, acting alone, would result in an electrically balanced liquid-crystalline structure. However, the A band lattice in living muscle appears to be a volume constrained liquid crystal (2,3). Thus, the role of osmotic compressive forces operating across the sarcolemma has been incorporated into the model.

The two populations of actin and myosin filaments vary in size as well as in ratio to each other depending upon the species of animal and the particular muscle in which they are observed. Most of the data to be discussed here have been obtained from the long tonic striated muscle fibers of the walking legs of crayfish (Orconectes). As in all striated muscles, the myosin filaments are alligned in a smectic "B" type lattice (hexagonal with the filaments in register). In this particular muscle the thick filaments, composed primarily of myosin, are approximately 4.4 um long and 18 nm in diameter. The

0-8412-0419-5/78/47-074-248$05.00/0

substructure of the thick filaments has not been ascertained, but appears to be composed of myosin molecules (M.W. 470,000; 150 nm long x 2 nm in diameter) arranged in an overlapping manner with a 14.3 nm displacement to form a helical rod with seemingly nine molecules per turn or a multiple thereof. The filaments have a net negative charge with the isoelectric point at approximately pH 4. In these crayfish muscle there is an approximate 60 nm interaxial spacing when stretched to the point where there is no thick-thin filament overlap. It should be emphasized that in these particular muscles there is no evidence of any structural connections between the myosin filaments.

The thin filaments, composed primarily of actin, are approximately 3 um long and 8 nm in diameter. The filament is basically formed by a nonintegral double helix of globular actin monomers (M.W. 50,000; 5.5 nm in diameter) which have a period of approximately 37 nm with approximately 13 actin monomers per turn per strand. The thin filaments originate at the Z line in the center of the I band and, while negatively charged, do not normally form a lattice except close to their attachment to the Z-line in which region the lattice is square.

The regions of myosin filaments and actin filaments alternate along the muscle fiber, forming respectively A-bands which are anisotropic and I-bands which are isotropic, thereby giving the muscle its striated appearance. Since I-bands are conveniently divided by a Z-line, two half I-bands and their mutually shared A-band comprise the basic repeat unit of striated muscle - the sarcomere (Figure 1). The actin filaments interdigitate with the myosin filaments and the amount of interdigitation is dependent upon the sarcomere length and, hence, the degree of stretch or shortening of the muscle fiber. At physiological sarcomere lengths, the thin (actin) filaments interdigitate with the thick (myosin) filaments of the A-band lattice to form a double (interleaving) smectic array. The arrangement of filaments in crayfish muscle is normally such that each myosin filament is surrounded by 12 actin filaments, providing a 6:1 unit cell for the A band lattice (Figure 2).

When the cell membrane of the muscle fiber is made permeable by glycerination or removed by microdissection, the osmotic phase boundary is eliminated and the filament lattice has direct access to the bathing medium. The lattice of the skinned fiber retains the smectic configuration. It was initially

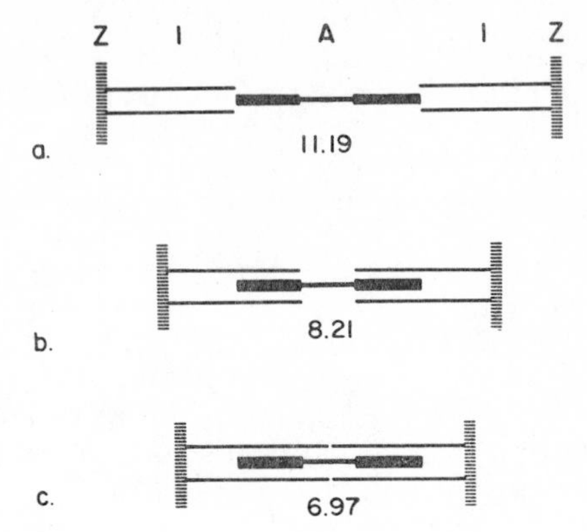

Figure 1. Schematic of the sarcomeric unit cell in crayfish striated muscle. The A-band, composed of myosin filaments, and the I-band, composed of actin filaments which originate at the Z-line, are indicated. The three sarcomere lengths depict muscle stretched to the extreme so that there is no filament overlap (a), normally stretched muscle (b), and shortened muscle (c), both of which indicate varying degrees of filament overlap.

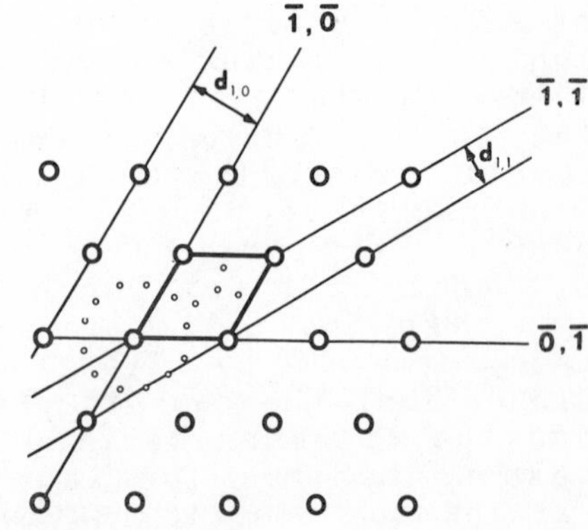

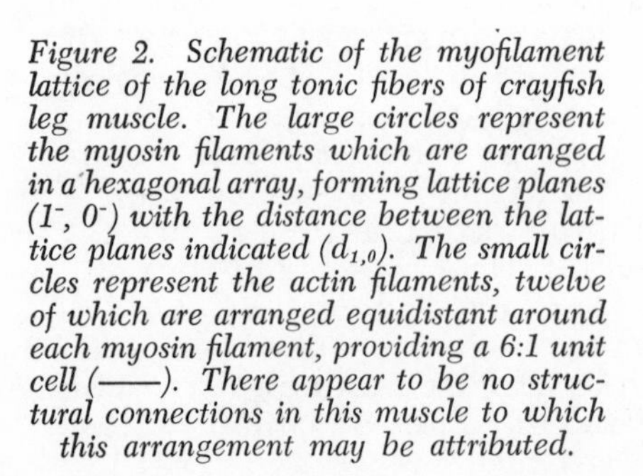

Figure 2. Schematic of the myofilament lattice of the long tonic fibers of crayfish leg muscle. The large circles represent the myosin filaments which are arranged in a hexagonal array, forming lattice planes $(1^-, 0^-)$ with the distance between the lattice planes indicated ($d_{1,0}$). The small circles represent the actin filaments, twelve of which are arranged equidistant around each myosin filament, providing a 6:1 unit cell (——). There appear to be no structural connections in this muscle to which this arrangement may be attributed.

hypothesized by Elliott (1) and Elliott and Rome (4) that the stability of the lattice was the result of a balance between van der Waal's and electrostatic forces. As Levine (5) had proposed for charged particles in an ionic medium, there is a relationship between the interaction energy and the distance with attraction at the longer distances and repulsion at shorter ones, between which is a point of equilibrium representing minimum interaction energy. Verwey and Overbeek (6) refined this relationship.

Since the long-range electrostatic forces are derived from the double layer associated with the surface of the filaments, the magnitude of these forces at any given distance is a function of any parameter which alters the effective charge on the filaments. This was observed in gels of Tobacco Mosaic virus by Bernal and Fankuchen (7) in which ionic strength and pH affected the interaxial spacing between virus particles in solution. They proposed that these parameters acted, respectively, by altering the degree of ionic screening between the particles and by altering the charge along the particles so that the magnitude of the repulsive forces varied.

Rome (8,9) and April, et al. (10,11) have shown by low-angle X-ray diffraction that glycerinated muscle and single skinned fibers, respectively, react in the predictable manner with changes in ionic strength, pH and divalent cation concentration. Based on the work of Rome (8,9) and the theory of Elliott and Rome (4), Miller and Woodhead-Galloway (12) calculated the expected effects of changes in ionic strength and pH on the lattice of thick filaments. Brenner and McQuarrie (13) formulated families of curves for the interaction energy between two cylindrical particles under various experimental conditions and pointed out that not only could a stable balance point be reached, but the stability would be greatly enhanced in a system in which each particle has six nearest neighbors. In skinned fibers the interaxial separation between thick filaments and the resultant skinned fiber diameter increases several times as ionic strength is decreased, indicating that the long-range electrostatic forces are exerted over greater distances as the ionic screening is decreased. This can be illustrated with data obtained from light microscopy and low-angle X-ray diffraction (Figure 3).

Similarly, as the pH of the medium bathing the skinned single muscle fiber is lowered, the interaxial spacing decreases to a minumum at approximately pH 4.4, then the spacing begins to increase. This represents the isoelectric point of the myosin filaments. Thus, as the charge on the surface of the filaments is varied with pH, the electrostatic repulsive forces are altered and the distance at which minimal interaction energy occurs likewise varies. This also can be well illustrated with data obtained from light microscopy and low-angle X-ray diffraction (Figure 4).

With respect to the double interleafing lattice that occurs with muscle shortening, Matsubara and Elliott (14) and April and Wong (15) have shown that interaxial separation is a linear function of sarcomere length. This can be explained on the basis of changing charge density within the A band as the negatively charged thin filament population interdigitates to various extents between the thick filaments. Since the total charge with the A band thereby varies with sarcomere length, the interaxial separation varies to maintain a state of minimum interaction energy. Equations have been formulated which concur with the experimental data (15,16).

The experimental data support the hypothesis that A band lattice of myosin filaments in the skinned fiber is an electrically balanced liquid crystal. However, in the presence of a viable cell membrane (sarcolemma) or an artificial phase boundary (such as polyvinylpirrolidone (M.Ws. 10,000, 40,000) in the skinned fiber preparation) the A band lattice behavior is markedly different and does not conform to the electically balanced condition (2,3). This is due to the presence of a Donnan-osmotic steady state which operates across the cell membrane and limits the volume of the muscle fiber.

The volume of the muscle fiber in physiological solution appears to be inadequate for the filaments to be at that distance which would allow an electrical balance between van der Waal's and electrostatic forces. The filaments in this volume-constrained condition are not free to redistribute to that distance which would result in minimal interaction energy. Osmotic studies with living intact single muscle fibers demonstrate that the fiber volume determines the interfilament separation (11). As the fiber volume is reduced, the filaments are confined to a smaller volume and the electrostatic repulsive forces must increase. Conversely, upon osmotic swelling, the filaments redistribute uniformly within the larger volume, utilizing the

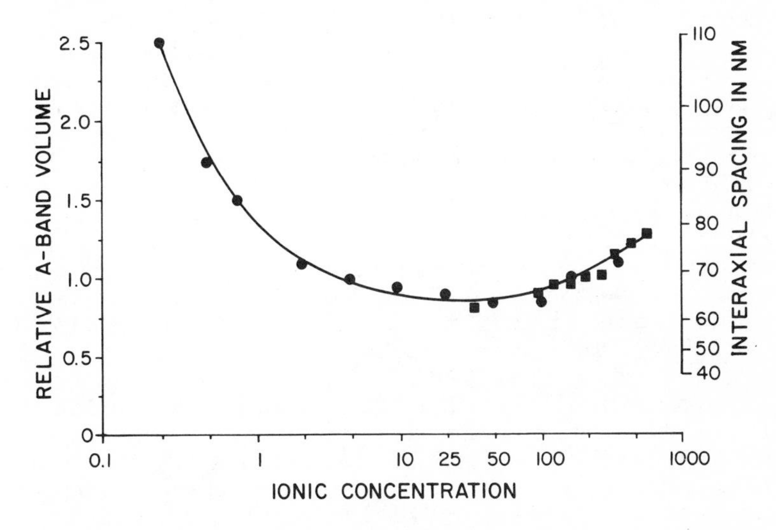

Figure 3. The effect of ionic strength on the filament lattice. Relatice A-band volume (determined by light microscopic measurements of skinned fiber diameter) and interaxial spacing (determined by low-angle x-ray diffraction of skinned fibers) plotted against the log of the ionic concentration of the bathing medium. The ionic strength was adjusted by varying the potassium propionate concentration of the medium which also contained 1 mM adenosine triphosphate, 1 mM magnesium chloride, and 10 mM ethyleneglycol-bis(B-aminoethyl ester)-N,N′-*tetraacetic acid (EGTA) to prevent contraction.*

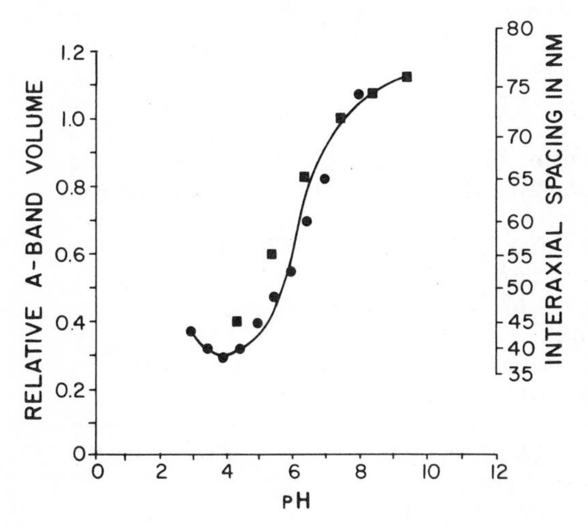

Figure 4. The effect of pH on filament spacing. Relatice A-band volume (determined by light microscopic measurement of skinned fiber diameter) and interaxial separation (determined by low-angle x-ray diffraction of skinned fibers) plotted against the pH of the bathing medium which contained 200 mM potassium propionate, 1 mM magnesium chloride, 10 mM EGTA with the pH adjusted with propionic acid.

interaction energy derived from the augmented electrostatic forces. A limit to lattice expansion occurs with a balance between van der Waal's and electrostatic forces. At this point the lattice changes from one of volume constraint to one of electrical balance (3). A similar volume-constrained or "non-equilibrium" liquid crystalline condition was described by Bernal and Fankuchen (7) in TMV gels. They observed that the interaxial separation between the virus particles was a function of virus concentration, i.e., available volume.

The concept of a volume-constrained liquid crystalline lattice has provided the answer to the long perplexing question in muscle physiolgy as to why the filament lattice varies as the reciprocal root of the sarcomere length during shortening. It is basically a matter of the volume of the A band varying reciprocally as the sarcomere length which is a function of filament overlap during shortening so that the filaments redistribute uniformly within the A-band as predicted in a volume-constrained liquid crystal.

While the nature of the forces affecting lattice stability have been identified by these studies, their precise magnitudes have not. It will be necessary to know these values to reconcile emperics with theory. Our work is now directed toward quantitating these forces. Employing polyvinylpyrrolidones as osmotic phase boundaries in the skinned fiber in an attempt to quantitate the osmotic forces and derive the electrostatic forces (17), it should be possible to estimate the electrostatic and van der Waal's forces fairly accurately.

What advantage does the liquid crystalline lattice lend to muscle and muscle function? The liquid crystalline characteristics provide the permeability necessary for biochemical reactions to occur within the A band lattice, i.e., calcium and ATP must diffuse into the lattice to activate the acto-myosin interaction and trigger the generation of tension. Concomitantly, the liquid crystalline characteristics provide and maintain structural integrity for the directional development of contractile force and transmission of the resultant tension. Finally, the system must be able to alter the degree of filament overlap and to change its shape during shortening or stretch while maintaining its structural integrity. Only a liquid crystal could provide all of these characteristics necessary for muscle function. (Supported by grants from NIH and MDAA.)

Literature cited:

1. Elliott, G.F. 1968. J. theor. Biol. 21:71-87.
2. April, E.W. 1975. Nature (London) 257:139-141.
3. April, E.W. 1975. J. Mechanochem. Cell Motility 3:111-121.
4. Elliott, G.F. and E. Rome. 1969. Mol. Crystals & Liquid Crystals 8:215-219.
5. Levine, S. 1939. Proc. Roy. Soc. (London) A170:165-182.
6. Verwey, E.J.W. and J.T.G. Overbeek. 1948. The Theory of the Stability of Lyophobic Colloids. Elsevier.
7. Bernal, J.D. and I. Fankuchen. 1941. J. Gen. Physiol. 25:111-165.
8. Rome, E. 1967. J. Mol. Biol. 27:591-602.
9. Rome, E. 1968. J. Mol. Biol. 37:331-344.
10. April, E.W., P.W. Brandt and G.F. Elliott. 1971. J. Cell Biol. 51:72-82.
11. April, E.W., P.W. Brandt and G.F. Elliott. 1972. J. Cell Biol. 53:53-65.
12. Miller, A. and J. Woodhead-Galloway. 1971. Nature (London) 229:470-473.
13. Brenner, S.L. and D.A. McQuarrie. 1973 J. theor. Biol. 39:343-361.
14. Matsubara, I. and G.F. Elliott. 1972. J. Mol. Biol. 72:657-669.
15. April, E.W. and D. Wong. 1976. J. Mol. Biol. 101:107-114.
16. Elliott, G.F. 1973. J. Mechanochem. Cell Motility 2:83-89.
17. April, E.W., M. Farrell and J. Schreder. 1977. Biophys. J. 17:174a-175a.

RECEIVED December 8, 1977.

INDEX

D

E

F